Frank Castle

An Elementary Course in Practical Physics

Frank Castle

An Elementary Course in Practical Physics

ISBN/EAN: 9783337277529

Hergestellt in Europa, USA, Kanada, Australien, Japan

Cover: Foto ©berggeist007 / pixelio.de

Weitere Bücher finden Sie auf **www.hansebooks.com**

AN ELEMENTARY COURSE

IN

PRACTICAL PHYSICS

BY

F. CASTLE, M.I.M.E.

*Mechanical Division, Royal College of Science; Lecturer on Mechanics, Geometry, etc.,
Morley College, London; Author of " Notes on Advanced and Honours
Theoretical Mechanics" in " The Practical Teacher"*

THOMAS NELSON AND SONS
London, Edinburgh, and New York
1899

PREFACE.

The older method of giving statistical information on science subjects was very unsatisfactory, alike to teacher and to student, and is now being replaced by a better system, in which the student, instead of seeing an experiment performed by the practised hand of a teacher, is furnished with the necessary apparatus, and with information sufficient to enable him to carry it out with some approach to accuracy.

In this manner, instead of attempting the hopeless task of remembering a law, often expressed in language which he cannot understand, he finds from careful experiments, and by the liberal use of squared paper, that such a law exists—he can even trace it in a rough experiment (and the apparatus which is sometimes given to students is as rough as it is possible to make it); but with care and proper precautions he is able to obtain either an accurate result, or at least a good approximation to it. Having proceeded so far, there is, at any rate, some incentive to induce him to try to understand more clearly what the law implies, and it thus becomes of vital interest and importance, instead of a mere jargon of words to be carefully remembered for examination purposes.

As will be seen on reference to the following pages, the attempt is made to include only those experiments which are well within the capabilities of first and second year students in science schools and classes.

The grouping together of one or more sections of each subject into chapters affords an opportunity of giving at the end of each a short recapitulatory summary, and, in addition, numerical exercises on the subject-matter. These are chiefly selected from the examination papers of the Science and Art Department. Solutions to these are given where necessary, and the answers to the remainder.

PREFACE.

It is an easy matter to assume that by means of a simple experiment an important law is understood; but if the knowledge so gained is not sufficient to solve a fairly easy numerical question, it indicates that the matter is not so clear as it ought to be. Hence these questions will be found useful, not as an end, but as a means to an end.

During an experience of over twenty years' teaching in science classes, the writer has found that the majority of students fail to realize the great importance of Physics, especially of Elementary Mensuration.

As will be seen on reference to the following pages, considerable space is devoted to this subject. The experiments, although numerous, are so arranged that they may be taken in any desired order, and, if necessary, at the discretion of the teacher the more difficult ones may be omitted altogether.

The same remarks apply to the sections dealing with Mechanics, Sound, Light, Heat, Magnetism, and Electricity.

In the few simple and practical experiments in Mechanics, the excellent method introduced by Professor Perry, D.Sc., F.R.S., which is now being generally adopted, is indicated, and consists in setting apart suitable apparatus for each experiment. This is easily made and fitted up, is always ready for use, and enables a student to obtain fairly accurate quantitative results.

In a book of this kind it would be impossible to acknowledge all the help received from various sources, but the writer is especially indebted to his brother, Mr. F. G. Castle, A.I.M.E., and to Mr. G. W. Fearnley, A.R.C.S., for many valuable suggestions and corrections in the sections dealing with Mensuration, Mechanics, and Heat; also to Mr. C. A. West, A.R.C.S., A.I.C., and Mr. J. Schofield, A.R.C.S., whose valuable aid in the sections dealing with Sound, Light, Magnetism, and Electricity has in no small measure contributed to any good feature which this part of the work presents.

LABORATORY WORK.

It should be carefully noted by the student that the educational value of the work done by him will depend entirely *on the way it is done*. The results will be valueless unless in each case an effort is made to ensure that the most accurate results have been obtained. When this is done, it will be possible to illustrate the general principles of science, to reason in a clear manner on the facts observed, and to obtain clear ideas as to the cause of any discrepancy between the observed values and those obtained from calculation. The following rules should be observed :—

1. Before commencing an experiment, see that all the materials required are in readiness and in good working order; read all about the experiment, so that before you begin you have a clear conception of what you are about to do.

2. Use the apparatus as directed in the instructions, and note briefly any inferences you can make as the work proceeds.

3. A methodical record of all work done should be made, and a notebook should be kept exclusively for laboratory work. This should be about six inches wide by eight inches long, with stiff covers. A smaller book may be used if desirable, but the larger one will be found to be much more convenient.

4. The number or description of the experiment must be entered also. Suitable columns, with the headings as shown in the following experiments, should be prepared, and all results obtained entered in their proper places.

5. The numbers recorded in any column must not be the result of calculations, but those obtained from the experiments.

6. The right-hand page may be kept for the record of work done, and the left-hand page for sketches of the apparatus used. These sketches should always be made, *not sketched from the book but from the apparatus actually used*.

7. When the experiment is completed, remove all apparatus not required for the next experiment; and before leaving the laboratory, see that all things are put away in their proper places.

8. After the experimental work is finished, proceed to make the necessary computations. All equations and results should be entered on the right-hand page, the arithmetical work as neatly as possible on the left-hand page, in such a position as not to interfere with the sketch previously made of the apparatus.

9. Wherever possible, the values obtained should be plotted on squared paper. This enables slight errors of observation to be readily corrected. The paper may be cut to a suitable size and inserted in the notebook (pasting a narrow strip along one edge for that purpose). In this manner all the results may be kept together.

10. The notebook must be handed in to the instructor for examination, and the corrections, if any, indicated should be made by the student before proceeding to the next experiment.

CONTENTS.

I.	PRELIMINARY CONSIDERATIONS,	9
II.	MENSURATION—MEASUREMENT OF LENGTH,	18
III.	MENSURATION—AREA AND VOLUME,	35
IV.	HYDROSTATICS,	52
V.	PROPERTIES OF AIR—BAROMETERS—BOYLE'S LAW,	71
VI.	REPRESENTATION AND MEASUREMENT OF FORCES—PARALLEL FORCES AND CENTRE OF GRAVITY,	77
VII.	WORK AND ENERGY—PRINCIPLE OF WORK—SIMPLE MACHINES,	92
VIII.	PENDULUM AND ATTWOOD'S MACHINE,	110
IX.	SOUND,	117
X.	VIBRATIONS OF A STRETCHED WIRE OR STRING MONOCHORD,	124
XI.	LIGHT,	130
XII.	REFLECTION—REFRACTION,	136
XIII.	MIRRORS—LENSES,	143
XIV.	HEAT—TEMPERATURE—EXPANSION OF SOLIDS,	152
XV.	EXPANSION OF LIQUIDS AND GASES,	157
XVI.	THERMOMETERS,	162
XVII.	SPECIFIC HEAT—CHANGE OF STATE,	168
XVIII.	TRANSMISSION OF HEAT—CONDUCTION—CONVECTION—RADIATION,	176

CONTENTS.

XIX.	MAGNETISM, ...	180
XX.	TERRESTRIAL MAGNETISM, ...	186
XXI.	ELECTRIFICATION BY FRICTION—POSITIVE AND NEGATIVE ELECTRIFICATION—CONDUCTORS AND INSULATORS, ...	191
XXII.	ELECTROSCOPE—ELECTRIC INDUCTION—PROOF PLANES —ELECTROPHORUS, ...	194
XXIII.	DIFFERENCE OF POTENTIAL—ELECTRIC DENSITY,	199
XXIV.	VOLTAIC ELECTRICITY, ...	203
XXV.	ELECTRO-MAGNETISM, ...	211

ANSWERS TO EXERCISES, ...	220
TABLE OF USEFUL CONSTANTS, ...	222
INDEX, ...	223

AN ELEMENTARY COURSE IN PRACTICAL PHYSICS.

CHAPTER I.

PRELIMINARY CONSIDERATIONS.

EVEN to the beginner, correct ideas of **space, time,** and **mass** are absolutely necessary, and the accurate measurement of these quantities is of the utmost importance. In mechanics, and in all sciences, we have to make measurements of lengths, areas, volumes, times, masses, or to estimate quantities of heat, electricity, etc. When very accurate measurements have to be made, carefully-constructed and costly apparatus is necessary; yet it is possible, with sufficient care, to obtain very good results with apparatus of a very simple character and costing but little. It also frequently happens that the principles involved in the use of an instrument are best illustrated by a simple type; which, unlike costly apparatus, is always ready for use. It is by such simple forms of apparatus that the following experiments are to be performed.

British and Metric Units.

In measurement, it is necessary to find some number indicating the magnitude of the quantity we are measuring; hence some definite quantity of the same kind must be selected as the **unit.** Thus we may express **Length** in **inches, feet,** or **miles,** or in **centimetres, metres,** or **kilometres**; **Area** in **square inches** or **square feet,** or in **square centimetres,** etc.; **Volume** in **cubic inches, cubic feet,** or in **cubic centimetres**; **Weight** or **mass** in **pounds** or **tons, grams** or **kilograms.**

Measurement of Length.—To measure length, a *unit* and a **standard** are necessary. The two units which are in use in this country and the United States of America are the **yard** and the **metre**.

The **English unit of length** is the **imperial yard**. This is defined by Act of Parliament to be the distance between two lines on a certain bronze bar, when the bar is at a temperature of 62° F. This bar is deposited in the Standards Department of the Board of Trade.

The measurement of a quantity is made by finding how many times the *unit*, which is a definite quantity of the same kind, is contained in it.

The two systems of units to which one yard and one metre respectively refer are—the **English**, and the **French** or **Metric**. The latter is chiefly used for scientific purposes.

The fundamental units of length, weight, and time respectively are:—

English—the **yard, pound,** and **second**. **Metric**—the **metre, kilogram,** and **second.**

From the fundamental what are called the *derived units* are obtained, by dividing the unit by suitable submultiples or multiples.

Derived Units.—The yard is divided into three equal parts, each of which is *a foot*. The foot is divided into twelve equal parts or *inches*, these being further subdivided into eight, or, better, into ten equal parts.

The foot and the inch are sometimes used for scientific purposes.

A larger multiple is the *mile*, equal to 5,280 feet or 1,760 yards.

F.P.S. System.—In the so-called absolute system of units in use in this country, the foot, the pound, and the second are used: this is known as the foot-pound-second system, abbreviated into *F.P.S. system.*

C.G.S. System.—In the metric or French system the centimetre, gram, and second are used, or the centimetre-gram-second system, and written *C.G.S. system.*

Area and Volume.—Units of area and volume depend directly on those of length. Thus the *unit of area* is the area of a square, whose side is the unit of length.

The unit of volume is the volume of a cube, whose edge is the unit of length.

PRELIMINARY CONSIDERATIONS.

The **English unit of area** is the area of a square whose side is one yard, and called the *square yard* (Fig. 1). For scientific purposes, the square foot and square inch, or the areas of squares whose sides are a foot and an inch respectively, are generally used.

The **unit of volume, or capacity**, is the *gallon*. The gallon is *the volume occupied by ten pounds of distilled water at a temperature of 62° F.*

A larger unit often used is the cubic foot. One cubic foot of water weighs nearly 1,000 oz., or, better, 62·3 lbs., when the temperature is 62° F. The weight of a cubic foot of water at temperatures 32° and 212° is 62·48 lbs. and 59·64 lbs. respectively. A pint of water weighs approximately a pound and a quarter.

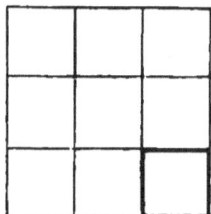

FIG. 1.—One yard and one foot; one square yard and one square foot.

Mass.—If we examine an iron ball and a wooden ball of the same size, the former is found to be much heavier than the latter. From this we infer that the masses or quantities of matter are different in the iron and the wood. Newton proved that the earth attracts all kinds of matter with the same force. Hence, if the masses were the same, the weights would be the same, for weight is the pull of the earth upon a body; but as this is not the case, there is more matter in the former than in the latter. Thus, for different substances, *mass is not proportional to volume, but is proportional to weight;* mass and weight are both proportional to volume when the substances are the same.

To obtain equal quantities of matter in the wood and the iron ball, the former would have to be much larger than the latter.

Definition.—The mass of a body is the quantity of matter in the body; it is measured by comparing it with the standard of mass.

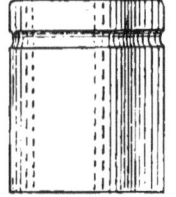

FIG. 2.—Standard pound: height, 1·35 inches; diameter, 1·15 inches.

The English standard of mass is the **pound avoirdupois** (Fig. 2), defined by the Weights and Measures Act, 1878, as the quantity of matter in a platinum cylinder deposited in the offices of the Board of Trade (Fig. 2). Copies of the standard, together with multiples and submultiples of it, are to be found in several parts of the country.

According to the Act, the comparison of any other mass with the standard unit is to be made by *weighing in vacuo.* The importance of this statement will be seen on reference to "Specific Gravity," p. 71.

Derived Units of Mass.—The multiples and submultiples in use are derived from this standard unit. Thus a ton = 2,240 pounds, or as written, 2,240 lbs.; also 1 lb. = 16 ounces = 7,000 grains.

Metric System.—In the French or metric system of weights and measures, *decimal* division—that is, division into ten or some multiple of ten—is exclusively used, and is much more convenient, both for measurement and calculation, than the method of dividing a yard into feet and inches.

In this system the unit of length is the **metre**. This length was originally defined to be the one-ten-millionth part of the length of a quadrant of the earth's meridian, from the pole to the equator; but further research has shown that this estimate is not accurate, and the *metre* may be best described as the distance between two marks on a metal bar deposited in the French Standards Office at Paris. Copies of the standard are to be found in various places in Europe.

Derived Units.—In the metric system all the measures of length, area, and volume are multiples or submultiples of the unit, and are obtained from the metre by multiplying or dividing by some power of 10. Thus the **metre** is subdivided into ten parts called **decimetres**, a decimetre into ten **centimetres**, and a centimetre into ten **millimetres**. When the distances to be measured are great, a thousand metres, called a **kilometre**, is used.

The following abbreviations are used:—

| Metre.................m. | Centimetre..............cm. |
| Decimetre.............dm. | Millimetre ....mm. |

It is often very convenient to write millimetres as tenths of a centimetre; thus a reading of 10 centimetres and 6 millimetres would be written as *10·6 cm.*

Comparison of Yard and Metre.—The *yard* and the *metre* have been carefully compared, and it is found that when both yard and metre are at a temperature of 62° F., a metre is 39·3704 inches. Hence a *decimetre* is 3·93704 inches, or very nearly equal to 4 inches.

Area and Volume.—The unit of area is the **square centimetre**, or the area of a square whose side is one centimetre.

The unit of volume is the **cubic centimetre**, or the volume of a cube whose edge is one centimetre.

If the edge of the cube be *10 centimetres*, or *1 decimetre*, the volume of the cube will be *1 cubic decimetre* (Fig. 3), and this is called a **litre**. Hence a litre contains *1,000 cubic centimetres.*

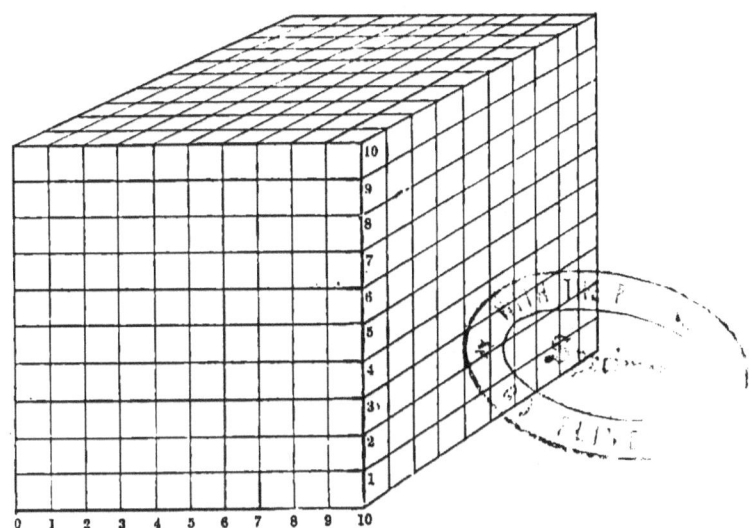

FIG. 3.—Centimetres. 1 decimetre = $\frac{1}{10}$ metre = 10 centimetres. Side of cube = 1 decimetre. Capacity, 1 litre. 1 cubic decimetre = 1000 cubic centimetres. Contains 1 kilogram of water = 1000 grams at 4° C.

The weight of pure water at 4° C. (the point of greatest density) which exactly fills the litre is called a *kilogram;* this is a little more than 2·2 lbs.

The kilogram is the unit of weight, or mass. It is divided into 1,000 parts, called **grams**. Thus the weight of a gram is $\frac{1}{1000}$th part that of a kilogram; or 1 c.c. of distilled water at 4° C. weighs exactly one gram, and would exactly fill a cube the length of the inner edge of which is 1 cm. in length.

Distilled water is very suitable for a standard; it is easily obtainable, and its density at the same temperature is always the same.

The metric standard of mass is called the **kilogram.** Originally it was so constructed as to have a mass equal to that of a litre of water at 4° C., but now it is defined to be the quantity of matter in a cylinder (Fig. 4) preserved at the Conservatoire des Arts et des Métiers, Paris. In comparing the French and

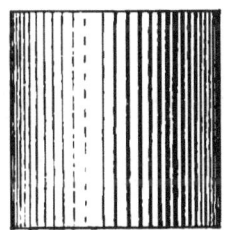

FIG. 4.—Platinum cylinder: 1 kilogram.

the British measures, we find 1 inch = 2·539977 centimetres; and similarly, a centimetre = ·3937043 inches.

It is often advisable to be able to convert British into metric measurements, or conversely. For this purpose the following approximate results are useful:—1 inch = 2·5 cm., but 2·5 = $\frac{10}{4}$; hence to convert inches to centimetres, multiply by 10 and divide by 4. To convert centimetres to inches, multiply by 4 and divide by 10.

COMPARISON OF BRITISH AND METRIC MEASURES.

Table I.

1 inch	= 2·539 = 2·5 centimetres approximately.
1 foot	= 30·48 centimetres.
1 millimetre	= ·039 inches = $\frac{4}{64}''$ nearly.
1 centimetre	= ·3937 inches = $\frac{4}{8}''$ nearly.
1 decimetre	= 3·937 inches = $3\frac{14}{15}''$ nearly.
1 metre	= 39·37 inches = 3·28 feet = 3' 3⅜" nearly.
1 square inch	= 6·45 square centimetres.
1 square foot	= 929 square centimetres.
1 cubic inch	= 16·39 cubic centimetres.
1 cubic foot	= 28,316 cubic centimetres = 1,728 cubic inches.
1 gallon	= 4·54 litres.
1 cubic centimetre	= ·061 cubic inch.
1 cubic metre	= 35·316 cubic feet = 1·308 cubic yards.
1 gram	= 15·43 grains = ·0353 ounces.
1 kilogram	= 35·27 ounces = 2·2 lbs.
1 grain	= ·0648 grams.
1 pound	= 453·59 grams.
1 litre	= 1·761 pints.

Properties and States of Matter.

Matter.—What is matter? To this question we cannot give a definite and conclusive answer. We refer to *wood, iron, air, water*, etc., as different kinds of matter, or as that which occupies space, and which we are able to recognize readily by one or other of our senses. The various kinds of matter are very different in their character: some are *hard*, others *soft;* some *fluid*, others *solid;* some *flexible*, others *brittle;* and so on. Each different kind of matter is called a *substance;* a limited portion of matter is known as a *body*. Perhaps the best definition of matter, and the one most suitable for our present purpose, is, **Matter is that which can be acted upon by, or which can exert, force.** All the bodies with which we have to deal consist of a definite quantity of matter. This matter acted upon by the pull of the earth, or **the force of attraction of the earth, gives the weight of a body.** What is called the mass of a

PRELIMINARY CONSIDERATIONS.

pound is a definite quantity of matter, but the weight of a pound is a definite amount of force. It is important to note that *the weight of a body measures its mass*, but the two terms, *mass* and *weight*, are names of two quite different things.

Properties of Matter.—We are familiar with matter in three states—**solid, liquid,** and **gaseous.** The term **fluid** is often used to denote the latter two states.

Some substances, such as water, may exist in all three states at different temperatures. Thus, *ice*, a solid, when heated becomes *water*, a liquid, and further heated becomes *steam*, a gas. **Permanent rigidity** may be said to be the distinguishing characteristic of a solid; unlike a liquid, which tends to yield either gradually or at once to any small force or forces tending to change its shape. A fluid *cannot offer permanent resistance to forces which tend to alter its shape.* Many substances which appear at first sight to be solid are found to be semi-solid or *plastic substances*. Thus lumps of pitch may be piled in a vessel, and apparently are solid masses; but if left for a short time under their own weight, or exposed to a slight rise in temperature, a viscous liquid is formed, and is found flowing over the sides of the vessel.

The principal properties of matter are:—
Extension, Impenetrability, Porosity, Compressibility, Cohesion, Divisibility, and **Inertia.**

Extension.—Every body occupies some space; the amount of this space is the extension of the body, which is in all possible directions, but all directions are included in the *three dimensions of length, breadth, and thickness.*

Porosity.—The constituent molecules of any body are separated by minute spaces or pores. Thus all bodies are more or less permeable to fluids. The spaces are visible in a sponge or in porous earthenware; but although invisible in water and alcohol, they can be shown by experiment to exist.

APPARATUS.—Glass tube (about 1 cm. bore by 90 cm. long, or $\frac{1}{2}$" bore × 36" long), one end sealed; water; methylated spirit.

EXPERIMENT 1.—Fill the tube about half full with water; now add methylated spirit until the tube is quite full; carefully close the open end of the tube either by placing the thumb over the open end or by using a tightly-fitting cork, and shake the two liquids together so that they mix. When this is done, a diminution in volume is found to have taken place: the two liquids when mixed occupy a smaller space than before, as is shown by the empty space at one end of the tube. That water is porous and will absorb air or gases, is shown by the presence of gas in a bottle of soda-water.

Impenetrability.—As two bodies cannot occupy the same space at the same time, there cannot be two particles occupying the same space at the same time. Hence impenetrability is that property of matter by which it occupies space to the exclusion of other matter. This statement seems to be contradicted by such substances as salt and sugar, which when placed in water are dissolved, and apparently disappear. What happens is, the particles of salt or sugar go into the minute spaces between the particles of water. They may be recovered by boiling off all the water.

Compressibility.—Since all matter is more or less porous, therefore under pressure a given mass (or quantity of matter) becomes smaller. Gases are very compressible; liquids are but slightly compressible. The compressibility of air, as in the air-gun or in a child's pop-gun, is well known. A cubic inch of air under a pressure of one atmosphere—a pressure equal to 14·7 pounds per square inch—would have its volume reduced one-half when under a pressure of two atmospheres—that is, its volume would be ·5 cubic inches. A cubic inch of water under the same conditions would only (fractionally) decrease its volume by $\frac{1}{20000}$th part.

Cohesion.—The molecules of a solid are held together by *cohesion*, an attractive force which exists and is exercised between similar molecules. The form in which matter exists depends upon the cohesion of its molecules. In solids the form cannot be altered unless the solid is subjected to considerable force. In liquids there is little or no cohesion, and the form can be changed with very slight force. Thus a liquid poured into a vessel takes the shape of the vessel.

Gases always tend to expand, and fill any vessel in which they are placed, exerting pressure against the sides. This is due to the repulsion, or absence of cohesion, between the molecules of the gas.

Divisibility.—If a small quantity of aniline red be reduced to fine powder in a mortar, the smallest quantity of this fine powder can be placed in a large quantity of water, and in this manner water can be obtained containing the millionth part of a grain of aniline red, as shown by the colour, so much diluted as to be just visible. Gold leaf can be obtained of a quarter of a millionth of an inch in thickness, and platinum wire of $\frac{3}{100000}$ inch diameter.

Inertia.—It is a matter of common experience that everything in the universe is apparently either at rest or in motion; also that matter resists external influences tending to alter its condition of

PRELIMINARY CONSIDERATIONS.

rest or of motion. It has no power to put itself in motion or to change its state in any way whatever. Any change must be produced by the action of some external force; also the effect of a given force in communicating or in changing the motion of a body will be found to depend on the size or quantity of matter in the body. This property of matter resisting change is called its *inertia*.

SUMMARY.

Units of Length, the yard and the metre.
Derived Units, foot, inch, centimetre, millimetre.
Area, square inch or square foot, square centimetre.
Volume, cubic inch or cubic foot, cubic centimetre, or cubic decimetre, called a litre.
Solids and Fluids, the two kinds of matter.
Solids, permanent and definite shape.
Fluids, divided into *liquids* and *gases*. *Liquids* have definite volume, but no fixed shape, and are but very slightly compressible, hence often called "incompressible" and "inelastic." *Gases*, compressible, having neither definite form nor definite shape, are elastic.
Mass, actual amount of matter in a body.
Weight, the force of gravity or the earth-pull.

Properties of Matter, etc.

Matter is that which occupies space and can be acted upon by, or which can exert, force.

Properties of Matter are *Extension, Impenetrability, Porosity, Compressibility, Cohesion, Divisibility,* and *Inertia*.

Matter Exists in Three Forms or States—*solid, liquid,* and *gaseous*. The same matter may, under suitable conditions, be made to assume all three states in succession.

Solids have a definite shape and volume; this shape and volume remain unchanged unless subjected to force.

Liquids have a definite volume but no definite shape. A liquid adapts itself to the shape of the containing vessel. Liquids are very slightly compressible, hence, the conditions remaining the same, the volume of a given quantity of liquid remains unaltered, however much its shape may vary.

Gases have neither definite form nor shape. They may be easily compressed, and expand indefinitely.

CHAPTER II.

MENSURATION—MEASUREMENT OF LENGTH.

In order to measure the length of objects, we must have some unit length divided into a number of equal parts, and by applying this unit to the objects we compare their lengths with regard to it, and thus with each other. A piece of wood or steel so divided is called a *scale* or *rule*.

Scales, or rules as they are sometimes called, are readily obtainable, and can be purchased for a small sum. Perhaps the scale most suitable and useful for our present purpose will be a steel straight-edge or rule divided into centimetres and millimetres on one edge, and into inches and either tenths or eighths of an inch on the other. Such a scale is shown in Fig. 5.

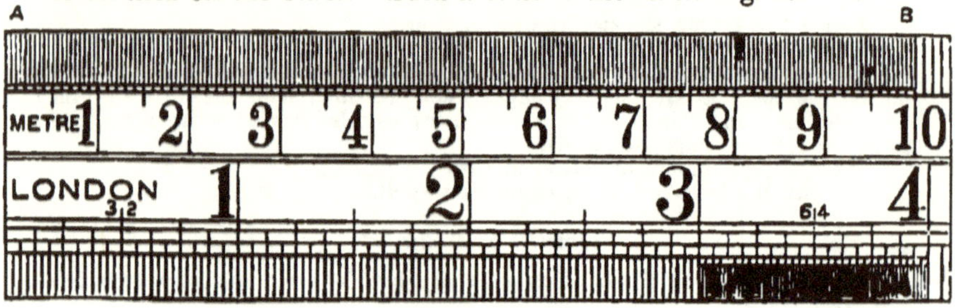

FIG. 5.—Steel straight-edge or rule. A decimetre divided into 10 centimetres and 100 millimetres.

In the **metric** system, the unit of length, as before stated, is the *metre*. This is divided into 10 equal parts, each part being called a decimetre. Such a length is shown at A B (Fig. 5), and is again divided into 10 equal parts, each called a centimetre (cm.). For smaller divisions the millimetre is used, being $\frac{1}{10}$th of a cm., and is shown by the 10 equal parts into which each centimetre is divided.

MENSURATION—MEASUREMENT OF LENGTH.

Instruments Used to Measure Length.—The longest lines, marked 1, 2......10, divide the *decimetre* (Fig. 5) AB into ten centimetres; each centimetre is divided into 10 equal parts, and therefore the distance between each small line on the scale is half a millimetre.

In the lower part of the scale the longest lines, 1, 2, 3, denote inches, the next in length half-inches; each half-inch is divided into 32 equal parts, enabling a reading to be made to the 64th of an inch; or the first inch or any other division on the scale may be divided into 50 or 100 parts, so that a reading, if care is exercised, can be made to the 50th or the 100th part of an inch. The divisions in the latter case are so close together that it is a difficult matter even with a lens to read with certainty.

Although it is practically impossible to measure a given length with absolute accuracy, in every case a certain degree of accuracy is requisite. Thus in a dimension which may include several feet, inches, and fractions of an inch, an error of $\frac{1}{10}$ or even $\frac{1}{2}$ inch may not be very important. In some measurements an error of $\frac{1}{10}$ inch, or $\frac{1}{10}$ of a millimetre, would render the result useless. It is important that the student should endeavour to make all measurements as accurately as possible. When it is required to transfer a dimension from the scale to paper, or when the scale is applied directly to ascertain the length of an object, care should be exercised that the measurement is made as carefully as possible. In the latter case the reading may be quite inaccurate, unless the graduated edge of the scale is brought as closely as possible into contact with the length to be measured, and the error due to parallax is avoided. (Parallax is due to a change in the position of the observer, causing an apparent change in the measurement.)

Two methods which may be used to measure the length, A B, of a block are shown in Fig. 6. If the scale is placed flat on the

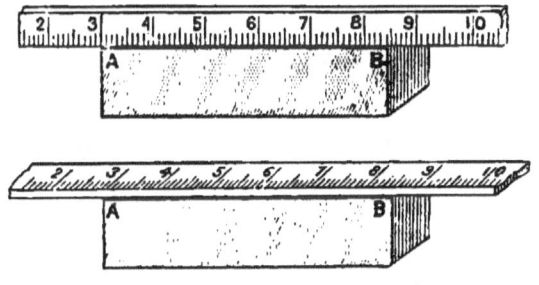

FIG. 6.—Two methods of using a scale.

surface of the block to be measured, the thickness of the scale may prevent an accurate reading being made, as the divisions on the scale are not in contact with the line to be measured. A much better way to use the scale is shown where, as in the upper part of the diagram, the scale is placed with its edge on the surface.

It is advisable not to use the end division to measure from, as it frequently happens that this division is slightly shorter than the others, or the end of the scale may not be cut exactly at right angles to its length. Even if correct at the outset, after being in use for a short time the end becomes worn, thus shortening the end division. This does not apply to boxwood and ivory scales, in which the zero of the scale does not fall on the end of the scale, but a little distance from it; in such scales the end division may be used.

To Copy a Scale.

APPARATUS.—Strips of wood about $\frac{3}{8}'' \times \frac{1}{2}'' \times 24''$ and $\frac{1}{8}'' \times 1'' \times 48''$ respectively; strips of cardboard; steel millimetre scale.

EXPERIMENT 2.—Bore a small hole, by means of a bradawl, about half an inch from the end of the strip; drive a stout pin through the hole, so that it projects about a quarter of an inch, and at the same distance from the other end cut a notch; fasten, by means of a piece of thread or fine string, a piece of blacklead pencil into the notch; also by means of drawing-pins fasten to the table the steel scale and a strip of cardboard in line with it, proceed to copy it, by placing the point of the pin in each consecutive division of the steel scale and drawing a corresponding line on the cardboard. Every fifth and tenth line should be longer than the others.

EXPERIMENT 3.—Make a metre scale to read to centimetres.

APPARATUS.—Some rectangular blocks of wood about $3'' \times 4'' \times 6''$, also one or two pieces of flat bar-iron 3 or 4 inches wide, and any convenient thickness; similar pieces of copper or glass plates, etc.

EXPERIMENT 4.—Ascertain the *average length* of each block supplied to you. To do this, measure the lengths (as accurately as possible) of the four edges; to ensure accuracy, the mean of several readings should be taken. Add the numbers so obtained, and divide by four to get the *average length*. In a similar manner obtain *average width* and *average thickness*. From the data so obtained calculate in each case the volume.

MENSURATION—MEASUREMENT OF LENGTH.

Tabulate as follows :—

Description.	Mean Length.	Mean Width.	Mean Thickness.	Volume.
...	.	...	...	...
...	...	...	...	...

In measuring a length by means of a scale, a difficulty occurs in reading with any degree of accuracy the fractional parts of a division. Thus, if, as shown in Fig. 7, a scale were applied to determine the length of an object—although, as already stated, each inch division of the scale could be mechanically divided into 100 equal parts, which would enable the reading to be made to the $\frac{1}{100}$ of an inch—the lines would be so close together that it would become difficult to distinguish between them; hence an open scale, such as that shown in Fig. 7, is preferred, and the length is seen to be

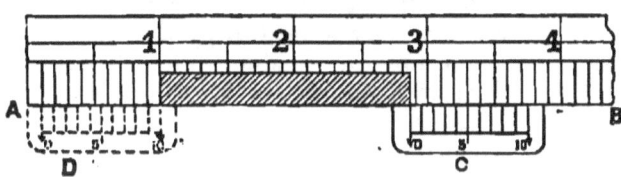

Fig. 7.—Length of an object. Use of vernier.

1·8 and a fraction. To obtain the fractional part, the distance between a division is then mentally divided into 10 equal parts, and the length is read as 1·85 inches. If the end of the cylinder does not quite reach to the middle of the division, then the length would be 1·83 or 1·84, according as the observer judged the distance to agree with the former or the latter; or if a little beyond the middle of the division, the reading would be 1·86 or 1·87. There is a certain amount of error introduced into this method; hence when accurate determinations of the fractional parts of a division are required, what is known as **a vernier** is used. The vernier is so very important in practical work, and especially in laboratory work, that it will be necessary to refer to it at some length.

The Vernier.—The vernier, which is made in many forms, supplies a ready means of measuring accurately the fractional part of a division.

One form is shown in Fig. 7, and consists of a graduated scale either in inches or centimetres, having another scale, or vernier, as it is called, which can slide freely along its graduated edge, A B. As will be seen from the vernier C, a length equal to 9 divisions on the scale A B is divided into 10 equal parts on the vernier. Each graduation then in the lower scale is less than each in the upper by $\frac{1}{10}$ inch; thus by using a vernier of the form shown, a reading to the $\frac{1}{100}$ of an inch can be made. To obtain the length of an object, the zero on the vernier is placed as shown in Fig. 7, so that it is coincident with the end of the cylinder, and the length is at once seen to be 1·8 and a fraction. To determine the fractional part, we look along the vernier and find that the 5th division is coincident with a division on the upper scale; hence the reading is 1·85 inch.

APPARATUS.—Small laths of wood, strips of cardboard, and steel millimetre scale.

EXPERIMENT 5.—Using a lath of wood or strip of cardboard, make a vernier to read to $\frac{1}{100}$ of an inch.

EXPERIMENT 6.—Make a centimetre scale reading millimetres, and a vernier to read to $\frac{1}{10}$ of a millimetre.

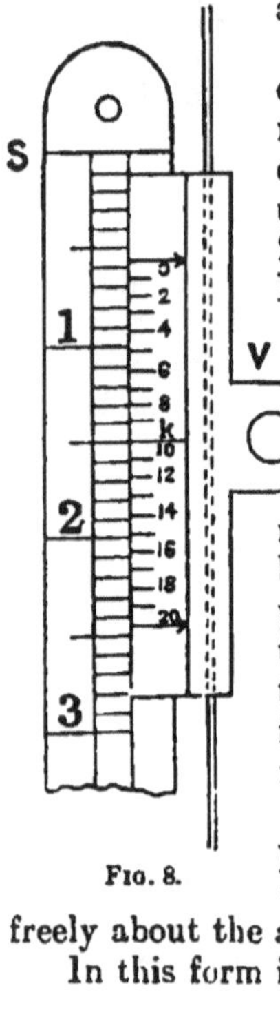

Instead of dividing a length equal to nine divisions into ten equal parts, a length equal to nineteen divisions may be divided into twenty equal parts. In this case (depending on the scale used) the fractional part can be estimated to the 200th of an inch or the 200th of a centimetre. In Fig. 8, a scale, S, may be fixed to any suitable frame or support, the vernier, V, being clamped to a length of wire which is being subjected to a tensile test. The zero on the vernier and the zero on the scale, S, were coincident at the beginning of the experiment with no load; when the wire is elongated by its applied load, and the vernier moves to the position shown, the elongation is seen to be between ·5 and ·6. Referring to the divisions on the vernier, we find at k the 10th division is coincident with a division on the scale; hence the reading is $·5 + \frac{10}{200} = ·55$.

FIG. 8.

For some purposes the graduated edge forms part of a divided circle; then the edge of the vernier will have the same curvature, and move freely about the axis of the circle.

In this form it is often used to determine the fractional parts

MENSURATION—MEASUREMENT OF LENGTH. 23

in the measurement of angles, but the principle is the same as already described.

Compasses, or Dividers.—A pair of compasses, or dividers, as shown in Fig. 9, may be used to transfer a given dimension, or to

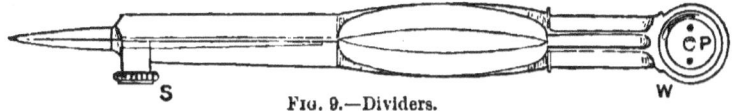

FIG. 9.—Dividers.

measure the distance between any two marks or lines. The two movable legs of the instrument are pivoted at P by a fairly tight joint, which is capable of adjustment by the screwed washer W, or, in the cheaper kinds, the joint is tightened by a blow from a hammer. (The dividers may or may not have a screw adjustment at S.) They are also useful where, on account of projections or cavities on the surface, it would be difficult, if not impossible, to apply a scale direct. When used to transfer a dimension from a scale, one of the points is placed in a division of the scale, and the dividers are opened until the other point is at or near the required dimension. A final fine adjustment can then be made by means of the screw. This may be rotated either forwards or backwards, giving to the point with which it is connected an outward or inward motion.

EXPERIMENT 7.—With your dividers make a scale a decimetre in length reading to millimetres, and use it to mark off a distance of 3·4 cm.

Diagonal Scale.—Fractional parts of a division can be obtained by means of what is known as a *diagonal scale.*

Such a scale, made of boxwood or ivory, is usually to be found along with others in a box of mathematical instruments. Unfortunately, the scale is too often quite inaccurate, and therefore almost useless. As shown in Fig. 10, the upper and lower edges are divided into ten equal divisions; then, by joining the zero on the lower to the division 1 on the upper line, we obtain a sloping line, to which the

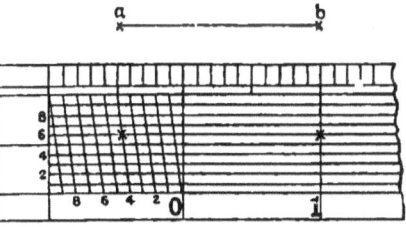

FIG. 10.—Diagonal scale.

remaining nine lines are drawn parallel. These sloping lines are intersected by horizontal lines 0......10, hence each of the equal upper divisions is $\frac{1}{10}$ inch, the division below is less by $\frac{1}{10}$ of a division; $\therefore \frac{1}{10}$ of $\frac{1}{10} = \frac{1}{100}$ inch; the next division below is less by $\frac{1}{10}$ of $\frac{1}{10} = \frac{2}{100}$, and so on.

Thus to estimate the lengths of a given line or the distance between two given points a and b (by using the dividers), we find that the distance extends a little beyond the 4th division, and, as shown, is coincident with the horizontal line through 6; hence the distance is 1·46 inches.

EXPERIMENT 8.—Make a diagonal scale to show centimetres, and to read to millimetres.

EXPERIMENT 9.—Make a diagonal scale to show inches, and to read to the hundredth of an inch.

Calipers.—In Figs. 11 and 12, what are known as outside and inside calipers are shown. These are useful when required to ascertain a dimension, such as the diameter of a cylinder or sphere, etc., to which an ordinary scale could not conveniently be applied. In using calipers, it must be remembered that the shortest distance across the cylinder is the diameter; unless a certain amount of care is exercised, a *longer* length than this will be obtained. If a scale be applied to the end of a cylinder, the measurement should be made across the centre, else a *shorter* length than the diameter will be obtained.

FIG. 11. FIG. 12.
Calipers.

Facility in using calipers is obtained after a little practice, the operator depending more on the sense of touch than on sight. When the diameter is obtained, the calipers are placed on a scale, and the dimension read off. To use the calipers, hold the cylinder or other object to be measured in the left hand, and open the calipers until the ends just clear. Next tap one of the outer limbs of the calipers until they will just pass over the object without exerting pressure to force them, then apply the calipers to a scale, and read off the dimension. In a similar manner the inside calipers may be used to ascertain the size of a cavity.

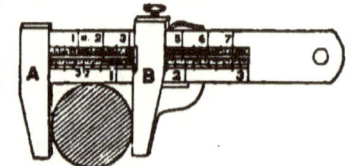

FIG. 13.—Slide-calipers.

Slide-Calipers.—In what are known as slide-calipers, shown in Fig. 13, the object to be measured is placed between the jaws

MENSURATION—MEASUREMENT OF LENGTH. 25

A and B; the jaw B is next moved up until in contact; then by means of the graduated scale shown, the dimension is known without having to use a separate scale, as required for the calipers. To ascertain the dimension with greater accuracy, a **vernier** may be attached to the slide-calipers.

Thus to determine the size of an object, such as a cylinder, with greater accuracy than can be obtained with the ordinary slide-calipers, a vernier may be attached to the instrument, as

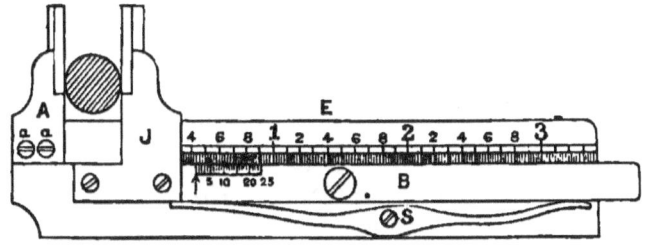

FIG. 14.—Slide-calipers with vernier.

shown in Fig. 14. In this instrument, by the use of the vernier attached, the dimension can be estimated to $\frac{1}{1000}$th of an inch.

By means of the two small screws, aa, the jaw, A, is fastened to the fixed scale, E. The movable jaw, J, is in like manner fastened to the sliding piece, B, and carries a vernier as shown, the vernier being kept in contact with the scale by the steel spring, S.

APPARATUS.—Several short lengths of brass, iron, and copper wire of different diameters; calipers, also slide-calipers.

EXPERIMENT 10.—Several short cylinders, brass, iron, and copper, of different diameters, are provided. Find the diameter of each by the outside calipers, and more accurately by the slide-calipers.

Wire-Gauge.—If the diameters of small cylinders of metal (or *wires*, as they are generally called) are required, then a special form of caliper is used, known as a *wire-gauge*. This is usually a flat piece of steel, with several openings in it round the edges; to each opening is given a number, instead of its width being marked in fractional parts of an inch. The diameters are then known by numbers instead of fractions. Such a wire-gauge may be circular, having the apertures at intervals along its circumference, or in the form shown in Fig. 15. Thus a piece of wire which would fit the aperture marked *d* (Fig. 15) would be described as No. 1, B. W.-G. (Birmingham wire-gauge).

When measured by the slide-calipers, the diameter is found to be 0·3 inch. It will be seen, on reference to Fig. 15, that as the numbers on the gauge increase, the size of the wire decreases; thus the

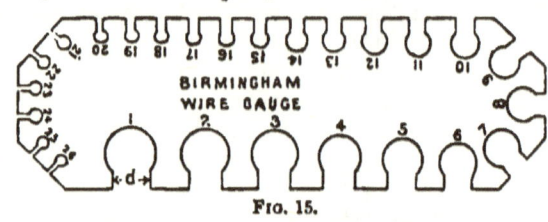

Fig. 15.

diameter of a piece which fits No. 17 is found to be 0·058 inch.

The diameters and areas of corresponding numbers are given in Table II.

APPARATUS as in Experiment 10.

EXPERIMENT 11.—Test the pieces of wire supplied by the wire-gauge and also by the slide-calipers, and compare your results with Table II.

Screw-Gauge.—This consists of a light frame, F, through the upper part of which passes a screw of fine pitch, rotated by a milled head, H, the rotation of H causing the sleeve, S, to advance or recede along the fixed cylindrical portion, A. When H is turned, the end of the screw, M, approaches the fixed end, N, and the scale at A is gradually covered by the sleeve. With rotation in the opposite direction the converse takes place.

The sloping edge of the sleeve, S, is divided into 50 divisions. If the pitch of the screw, M, be half a millimetre, it follows that if

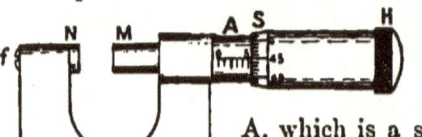

Fig. 16.
Screw-gauge.

H makes one turn, M will advance or recede through that distance. This distance is recorded by the scale on A, which is a scale reading to a millimetre: thus, as shown in Fig. 16, the distance between M and N is 5·45 millimetres; if the head, H, be rotated in watch-hand direction until the zero on the sleeve, S, is coincident with the division 5, the distance will be reduced to 5 millimetres, and the scale A would record that amount.

When M and N are just brought into contact, the zero on the sleeve should exactly cover the zero on the scale. To adjust the instrument, a small screw-driver inserted at *f* will cause N to move forwards or backwards as required. As one rotation of

MENSURATION—MEASUREMENT OF LENGTH.

the sleeve will increase or diminish the distance between M and N by half a millimetre, and as there are fifty divisions on the sleeve, a rotation through one division will cause M to move $\frac{1}{50} \times \frac{1}{2} = \frac{1}{100}$ mm.; hence it is possible to measure to the $\frac{1}{100}$th of a millimetre.

To use the instrument, first bring the ends M and N into such a position that they lightly grip an object placed between them; note the reading on the scale, and the fractional part on the sleeve S.

Care must be taken not to force the ends tightly together, or the screw thread may be injured.

The student should carefully examine, and see how "backlash" is prevented in a screw-gauge, and compare the arrangement with any ordinary bolt and nut, or any common application of a screw, such as a screw-press. In these applications it will be found that the screw can be rotated a short distance to the right or left without causing the nut at the end of the screw to advance. This "lost" motion is called "backlash." If the screw-gauge be examined, the arrangement by which the nut is divided and held together by a taper-nut is well worth attention. A sketch should be made in the note-book.

APPARATUS.—Several short lengths of wire of different diameters; calipers; screw-gauge.

EXPERIMENT 12.—Several pieces of wire are provided; estimate roughly the diameter of each. (*a*) Find the corresponding number on wire-gauge, and the diameter from Table II.; (*b*) find the diameter by the calipers; (*c*) and more accurately by the screw-gauge. Tabulate as follows:—

No. on B. W.-G.	Diameter obtained by Calipers.	Diameter by Screw-Gauge.

TABLE II.—BIRMINGHAM WIRE-GAUGE.

No.	1	2	3	4	5	6	7	8	9	10	11	12	13
inches.	·300	·284	·259	·238	·220	·203	·180	·165	·148	·134	·120	·109	·095
	14	15	16	17	18	19	20	21	22	23	24	25	26
	·083	·072	·065	·058	·049	·042	·035	·032	·028	·025	·022	·020	·018

Measurement of Length.

APPARATUS.—Three or four wooden discs, about an inch thick, of different diameters, some thin paper, and a scale.

EXPERIMENT 13.—**To find the circumference of a circle, and to verify the formula, circumference = π × diameter.** Measure the diameter of a disc by the scale or the calipers.

(a) Wrap round the disc, as tightly as possible, a strip of thin paper; then with a pin or the point of the dividers make two holes where the ends of the strip come together. Unroll the paper, and carefully measure the length between the holes so obtained. Do this for each disc. Tabulate as follows:—

Diameter.	Circumference.	$\dfrac{\text{Circumference}}{\text{Diameter}}$

In each case the number obtained in the last column will be 3 and a fraction. The fractional part will depend on the accuracy with which the numbers are obtained, but will be found in all cases to be not far from 3·14; hence the length of the circumference of any circle contains the length of the diameter 3·14 times.

This number denotes *the constant ratio of the circumference of a circle to its diameter*, and is usually represented by the Greek letter π (pronounced *pi*).

The value of π to four places of decimals is 3·1416; the number $3\tfrac{1}{7}$ is often used for convenience.

(b) Make a mark on the edge of the disc; put this mark on a division of the scale, and roll the disc along the scale (taking care that slipping does not occur) until the mark again touches the scale. Note the distance travelled, and tabulate as before. (c) Or make a mark on the disc and on the paper; put the marks together, and roll the disc until the mark again touches the paper; measure the distance between the first mark and the second, and proceed as before.

EXERCISES.

1. The diameter of a circle is 2 feet 4 inches; find the circumference.
2. The circumference of a circle is 31·416 inches; find its diameter.
3. The diameter of a circle is 6·4 cm.; find its circumference.
4. The circumference of a circle is 30 cm.; find the diameter.
5. How many turns will a bicycle wheel 26 inches diameter make in going 10 times round a cycling ground of 500 feet diameter? How many turns would the wheel make in a journey of 10 miles?

Diameter of a Sphere.—It would be very difficult, if not impossible, to measure the diameter of a sphere by means of a scale; but by means of calipers, or by two rectangular blocks of wood and a scale, it can be determined. Care must be taken that the ends of the blocks are at right angles to the sides. Thus in Fig. 17 the two blocks appear to touch along the whole surface; but if A be turned over, so that the top face comes to the bottom, as shown in Fig. 18, the blocks are seen to be inaccurate.

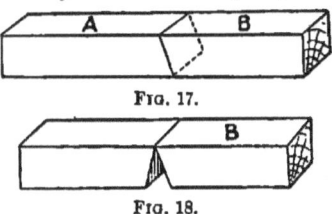

FIG. 17.

FIG. 18.

To measure the diameter of a sphere by two blocks (Fig. 19), having ascertained by testing, as explained in Figs. 17 and 18, that the ends of the blocks are at right angles to the faces, and therefore parallel to each other, the sphere is placed on any horizontal surface (such as the top of a table), the ends of the blocks are brought into contact with it, and the distance apart, which is the diameter of the sphere, can be measured by a scale as shown. To get the faces of the blocks in a straight line, they can conveniently be placed in contact with a third block of wood of any convenient dimension, or along a straight edge, such as the edge of the scale.

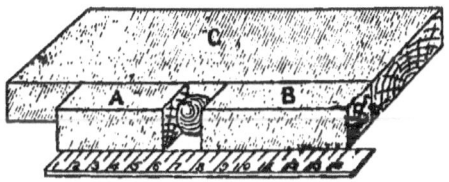

FIG. 19.—Measuring the diameter of a sphere.

APPARATUS.—Two blocks of wood with square ends, 3 or 4 inches edge, length 5 or 6 inches; wood or metal sphere, 3 or 4 inches diameter; a few bicycle balls; scale; calipers, and screw-gauge.

30 MENSURATION—MEASUREMENT OF LENGTH.

EXPERIMENT 14.—Measure diameter of sphere (*a*) by calipers; (*b*) by using the blocks, as shown in Fig. 19; (*c*) place the bicycle balls in a row, touching each other; measure the sum of the diameters, and dividing by the number used, obtain the diameter of one ball; test by using screw-gauge.

The Screw.—If a triangle A B C (Fig. 20), having its base equal to the circumference of the cylinder D E, be wrapped

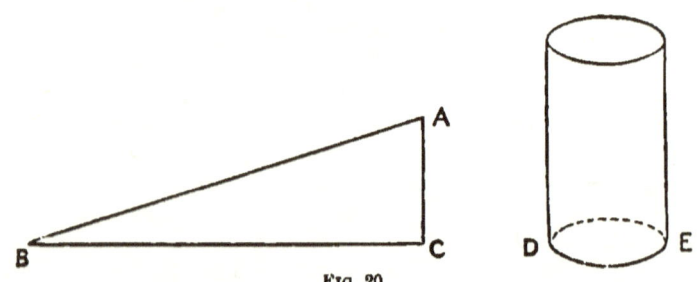

FIG. 20.

round it as shown (Fig. 21), and in such a manner that the point B is made to coincide with C, the edge A B will form a spiral line or curve round the cylinder.

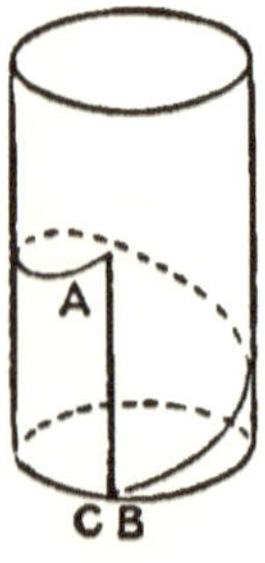

APPARATUS.—A cylinder 2 or 3 inches diameter; sheet of thin paper; drawing instruments.

EXPERIMENT 15.—Make a right-angled triangle A C B such that the base B C = circumference of the cylinder (Fig. 20). Cut out the triangle, and wrap it round the cylinder (Fig. 21) so that the line C A is at right angles to the base of the cylinder, and B coincides with C. Then it will be found that A B makes a spiral line or curve round the cylinder, and starting from A, in travelling once round the cylinder along the spiral line, we shall travel down the cylinder through a distance A C.

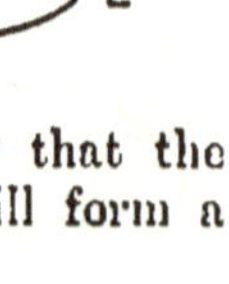

FIG. 21.

If we assume a projecting thread fixed to the outside of the cylinder, and coinciding with A B, as shown in Fig. 22, we obtain what is called a *screw thread*, and the distance A C is then called *the pitch of the screw*. If B C be of sufficient length, the line A B will make several turns round the cylinder, the pitch being $\frac{\text{length A C}}{\text{number of turns}}$.

FIG. 22.

The distance A C in any screw is usually either accurately known or can be determined by applying a scale; hence it is possible to

MENSURATION—MEASUREMENT OF LENGTH.

move in the direction of the axis of the cylinder any desired fractional amount. Thus a movement along the curve A B equal to one-fourth A B would correspond with a distance parallel to the axis of the cylinder equal to one-fourth the pitch.

A hollow cylinder, having a groove cut on its internal surface of such a size and shape that it just fits the screw thread referred to, is called a "*nut.*"

What are known as the square and the V threads are in general use. Fig. 23 shows a portion of a bolt with a square thread and a section through the nut, and Fig. 24 a V-thread bolt and nut. It will be noticed that the *pitch* is comparatively small. If

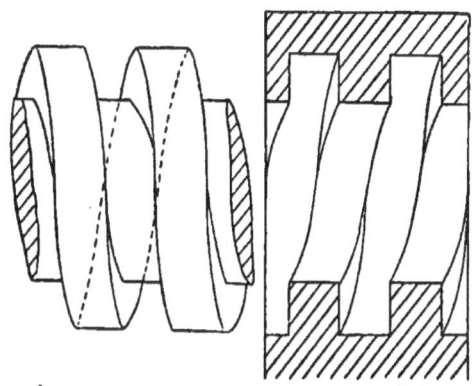

FIG. 23.—Square thread and section of nut.

the bolt B (Fig. 24) be prevented from rotating, and the nut N make one revolution, the distance between the inner edge of the nut and bolt-head is increased or diminished by a distance equal to the pitch of the screw. In a similar manner, any fractional part of the pitch is obtained by rotating the bolt through the corresponding part of a revolution.

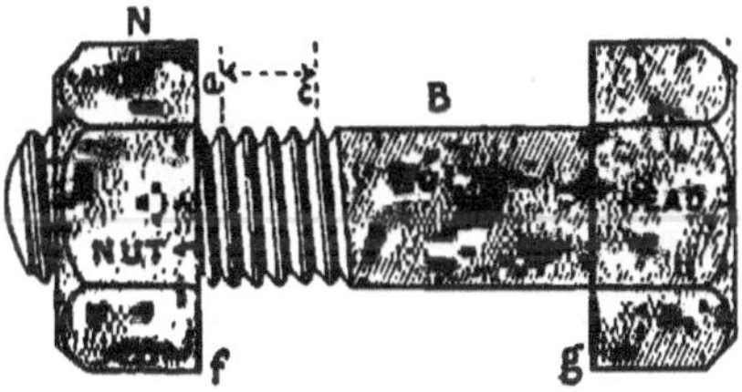

FIG. 24.—V-thread bolt and nut.

APPARATUS. — Several bolts of different diameters— say, $\frac{1}{2}''$, $\frac{3}{4}''$, $1''$, etc.

EXPERIMENT 16.—Determine and write down the *pitch* of the screw thread for each bolt supplied to you. To do this, count the number of threads (or ridges) in a measured length (ac, Fig. 24); then pitch $= \dfrac{\text{distance } ac}{\text{number of ridges between } a \text{ and } c}$.

Verify your calculation by rotating the nut a definite number of turns, and measuring the increase or diminution of the distance between inner edges of nut and bolt-head.

$$\text{Pitch} = \dfrac{\text{distance moved}}{\text{number of turns}}.$$

MENSURATION—MEASUREMENT OF LENGTH.

Measurement of Curved Lines.—The length of a curved line can be obtained with sufficient accuracy by the method shown in

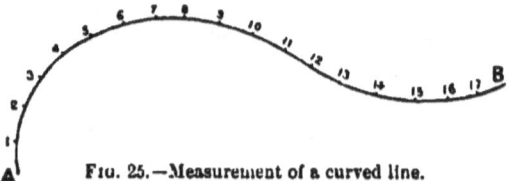

Fig. 25.—Measurement of a curved line.

Fig. 25. This consists in dividing the curve into a number of equal chords—A1, 12, 34, ... 17 B —as shown (the distances may be 2 or 3 mm., or ·2" or ·3" apart). If, as shown in Fig. 25, 18 divisions are required, and the distance between each is ·3", the length A B is $\dfrac{18 \times 3}{10} = 5\cdot4"$ approximately; the error involved is the substitution of a straight for a curved line, and the distance obtained is too short. A better result is obtained by using a piece of thread or fine string, as shown in Fig. 26. One end of the thread is made to coincide with point A, and a small portion of the thread to correspond with a short length of the curve. By shifting the position of the fingers another short length is added, until the length of the thread A′C represents the length of the curve up to C. By repeat-

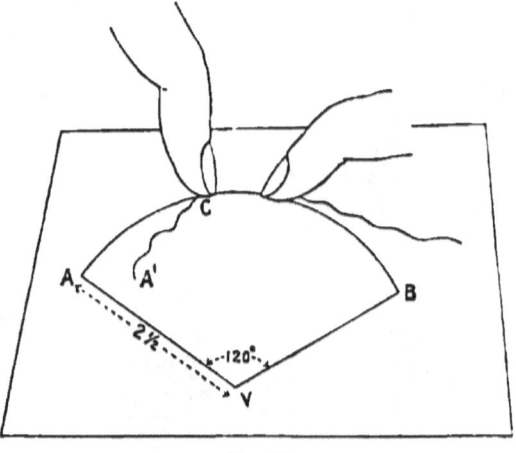

Fig. 26.

ing the process the whole length of the curved portion may be obtained, and measured by laying the used part of the thread on a scale.

EXPERIMENT 17.—The development of a cone is shown in Fig. 26. Determine and write down the diameter of the base and the height of the cone. (The height can be determined by taking the slant side and the radius of base as two sides of a right-angled triangle; the third side is the height required.)

EXPERIMENT 18.—Draw a circle 3·2 cm. radius, and use the two methods shown to find its circumference.

MENSURATION—MEASUREMENT OF LENGTH.

APPARATUS.—Wood cone, base 3 or 4 inches diameter, height 5 or 6 inches; pencil, compasses and scale, and thread.

EXPERIMENT 19.—Draw two lines, one on the slant side of a cone from the vertex V to a point A on the edge of the base, the other from A to the centre of the base; also draw a line on a sheet of paper equal in length to V A. Put the line V A on the cone in contact with the line on the paper on a flat surface, such as the top of the table; roll the cone until the point A again touches the paper; call this point B. With your compasses—V A or V B as radius, centre V—describe an arc of a circle, A B. Join V to A and B; the sector V A B will be the development of the surface of the cone as in Fig. 26.

EXPERIMENT 20.—Find the length of the curved line A B, and divide by $3\frac{1}{7}$; the number so obtained will be the diameter of the base. This can be verified by measuring the diameter by the scale.

Measurement of Angles.—When one line, B C (Fig. 27), intersects another, A C, at point C, the opening between the lines is called the angle B C A, or, if only one angle is formed at C, simply the angle C.

Fig. 27.

A length such as the distance between two points can be estimated as so many times the unit (feet or metres), but this is not suitable for *angular measurement*. In this case a convenient unit of measurement is a *degree*. In Experiment 13, the circumference of any circle was found to be 3·1416 times the diameter. If two points be taken on the circumference of a circle so that the length of arc between them is $\frac{1}{360}$ part of the circumference, the angle subtended by this arc at the centre of the circle is called a *degree*. In Fig. 27, if A C be produced to any point E, and D be drawn perpendicular to A E, then the angles A C D and D C E are equal to each other, each being a right angle. This is divided into ninety equal parts, each called a *degree*; each degree is further subdivided into sixty equal parts called *minutes*, and each minute again into sixty equal parts called *seconds*.

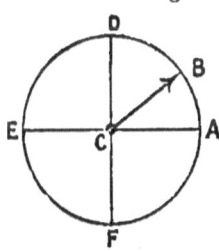

Fig. 28.

If, with C as centre, and with any radius such as C A, a circle be described, and the two diameters A E and D F be drawn at right angles to each other, the circle is divided into four equal parts, as shown in Fig. 28, each angle, A C D, D C E, etc., being 90 degrees; hence the circle represents 360 degrees, or 21,600 minutes.

It is convenient to write degrees thus °, minutes thus ′, and seconds thus ″.

If we assume C B to be a pointer or finger centred at C, and to be made to move round the circle in the opposite direction to the hands of a clock (this being the positive direction in which angles are measured), then when B is at A, the angle which C B makes with C A will be zero, and written 0°; when B is at D, the angle is 90°; when at E, is 180°; at F is 270°, and at A is 360°.

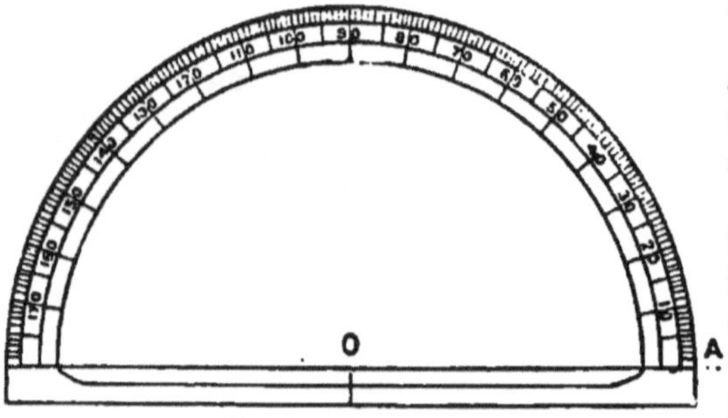

When it is required to set out or to measure an angle, a **protractor** is used.

Protractors, two forms of which are shown in Figs. 29 and 30 (for reading or marking off degrees), are made of various materials; but horn, boxwood, ivory, and metal are those generally used.

FIG. 29.—Protractor.

In each case, to measure an angle such as the angle ACB (Fig. 27), the line 0 A would be placed on CA, with 0 on C. The division on the protractor coincident with line C B would indicate the magnitude of the angle.

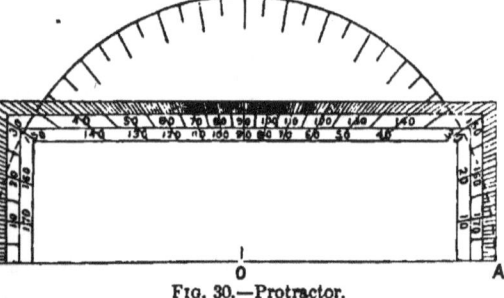

FIG. 30.—Protractor.

Conversely, to set out an angle of any given number of degrees, a line such as C A is drawn (Fig. 27), and the edge 0 A applied to it, so that 0 is on C, the edge 0 A coinciding with C A; the division indicating the number of degrees required is marked off and joined to C.

EXERCISE.

Set out a protractor, Fig. 29 or Fig. 30, on a strip of celluloid, horn, or stiff paper, and use it to set out angles of 25°, 30°, 45°, 120°, and 135°. Test the last four named angles with a set square.

CHAPTER III.

MENSURATION—AREA AND VOLUME.

Measurement of Area.—In measuring areas of surfaces, plane or otherwise, the unit used is the area of a square, each side of which is unit length; this unit length may be an inch, a centimetre, a foot, etc.

Area of a Square, and to verify the formula $area = a^2$.— If A B C D (Fig. 31) represent a square, each side of which is 4 units in length, then dividing A B and B C each into four equal parts and drawing lines through these points parallel to A B and B C respectively, the square is seen to be divided into 16 unit squares. Hence the area is 16 square units. Thus *the area of a square* $= a \times a = a^2$, where a denotes the number of units of length of a side or edge of the square.

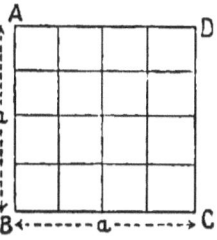

FIG. 31.—Area of a square.

Area of a Rectangle, and to verify $area = a \times b$.—Let A B C D (Fig. 32) be a rectangle, sides being 5 inches × 2 inches; dividing A D into five equal parts and A B into two, and drawing lines as before, parallel to A D and A B, we get 10 unit squares or 10 square inches. Hence the area of a rectangle is equal to **the product of two adjacent sides** $= a \times b$.

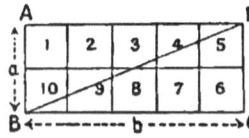

FIG. 32.—Area of a rectangle and triangle.

Area of a Triangle.—If B be joined to D, the figure is divided into two equal triangles; the area of each is therefore 5 square inches, or half the area of the rectangle, hence the area of each

$$\text{triangle} = \frac{\text{base} \times \text{altitude}}{2} = \frac{a \times b}{2}.$$

The result is evidently true

when the triangle is what is known as a right-angled triangle. We shall see that it is also true for any triangle.

When the numbers which represent the lengths of the adjacent sides of a rectangle are not whole numbers, the rule can easily be verified. Let H E (Fig. 33) be a rectangle, having its sides H B and B E equal to 2·5 and 5·4 cm. respectively. Mark off 5 cm. from B to D, and 2 cm. from B to A; through these points draw lines parallel to the sides of the rectangle; then area of rectangle A D is 10 square cm.

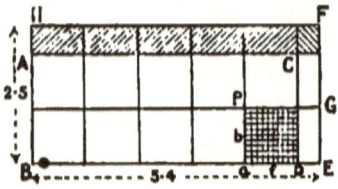

Fig. 33.—Area of a rectangle.

The shaded rectangles on the upper edge have each one side 1 cm., the other side ·5 cm., and as each, if placed on a square of 1 cm. side, would exactly cover one-half of it, the area of each is $\frac{1}{2}$ sq. cm.; hence the area of the five rectangles from H to C is $5 \times ·5 = 2·5$ sq. cm.

The two rectangles D G and G C have each one side 1 cm., the other ·4 cm.; if put on a sq. cm. side by side, they would cover exactly ·8 of it—that is, the area of D G and G C is ·8 sq. cm.

Finally, if we take a sq. cm. and divide each of the two sides into ten equal parts as shown at D P (Fig. 33), make ab equal to 5 and af equal to 4 of these divisions, then area of bf is $\frac{1}{5}$ of a sq. cm., or ·2 sq. cm.

Hence total area $= 10 + 2·5 + ·8 + ·2 = 13·5$ sq. cm. We should have obtained the same result by multiplying 2·5 by 5·4: $\therefore$ $2·5 \times 5·4 = 13·5$ sq. cm.; hence the rule is verified. Any other numbers can be selected, and the verification obtained in a similar manner.

EXPERIMENT 21.—Measure the length and width of a page of your note-book and find its area; also find the area of the top of your work-table or the top of your instrument case.

EXERCISES.

1. Given that a metre equals 3·2809 feet, find the number of square metres in 1000 square yards.
2. Find the length of the diagonal of a square whose area equals that of a rectangle of sides 16 and 10 feet long respectively.
3. Draw *on squared paper* a triangle having sides 5, 4, and 3 units respectively; the angle opposite the longest side will be found to be a right angle. Prove that the area of the square on the side 5 units long is equal to the sum of the areas on the other two sides.

Area of a Parallelogram.

To find the **area of a parallelogram**, and to prove that *the area* = **a** × **b**, where a is the altitude and b the base of the parallelogram. — In any parallelogram, A B C E (Fig. 34), let C D and B G be drawn perpendicular to B C; produce E A to meet perpendicular from B at G; then the triangle G B A has the three sides G B, B A, and A G equal to the three sides D C, C E, and E D respectively of the triangle D C E; hence the two triangles G B A and D C E are equal in all respects, by Euclid, Book I., Prop. 4 and 8.

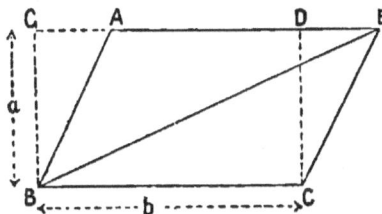

FIG. 34.—Area of a parallelogram.

Now if we take away the triangle D C E from A B C E and add instead the equal triangle G B A, we obtain a rectangle G B C D, and the area of the parallelogram is equal to the area of the rectangle. The result also follows at once from Euclid I. 35; ∴ **the area is base × altitude = a × b.** It follows that the area of the triangle B E C, which is half the parallelogram, is $\left(\dfrac{\text{base} \times \text{altitude}}{2}\right)$, and by Euclid, Book I., Prop. 41, each triangle is half the parallelogram. Draw the parallelogram on a sheet of cardboard, cut out the triangle B E C, place the triangle B E C in one pan of a balance, and the remaining half of the parallelogram in the other pan: the balance will be in equilibrium, showing that the weights are equal.

Use of Squared Paper.— A piece of paper, as will be seen on reference to Fig. 35, is divided by two series of parallel lines at right angles to each other. The distance between any two consecutive lines is usually $\frac{1}{10}$ inch; thus the sheet of paper is divided into a large number of little squares, the area of each

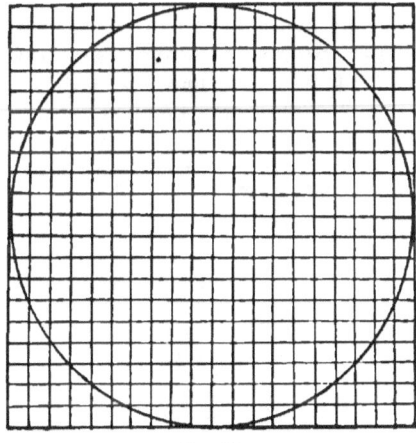

FIG. 35.

being $(\frac{1}{10})^2 = 0.01$ sq. inch. In some cases every fifth or tenth line is a different colour from the rest, enabling the number of squares in any given space to be easily calculated.

Area of a Circle.

APPARATUS.—Pencil compasses; squared paper; cardboard.

EXPERIMENT 22.—To find the **area of a circle** and to **verify the formula** $area = \pi \times (rad)^2$. (*a*) Draw a circle with radius $1\frac{1}{2}''$. It will be found that the circle encloses a number of complete squares and squares intersected by the circle. Estimate for those intersected the total area which lies within the circle, and so obtain the area. This will be found to be about 7 square inches. Proceed in a similar manner for circles of 2 inches and 4 inches in diameter.

Tabulate as follows :—

Radius.	Area.	Area / (Rad)²
1"	......	......
1½"	......	......
2"	......	......

The numbers obtained in the last column will be found to be either constant or nearly so, and each equal to $\pi = 3.14$. Instead of 3·14, its equivalent, $3\frac{1}{7}$, may be used. Thus the area of a circle $= \pi r^2$. This may be shown by drawing two circles on a sheet of cardboard. One is cut out as carefully as possible, and the other is circumscribed by a square as shown in Fig. 36. By drawing two diameters, as shown, four squares are obtained, the area of each being r^2. If now Oa be made equal to $\frac{1}{7}OA$, and ad be drawn parallel to OD, the shaded portion is seen to be $3r^2 + \frac{1}{7}r^2$ or $3\frac{1}{7}r^2$.

Cut out the shaded portion and verify that the weight, and therefore the area, is equal to that of the circle of the same size cut from the same sheet of cardboard.

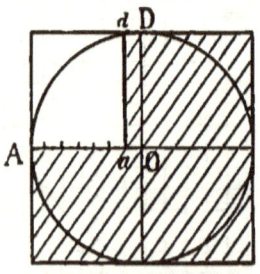

FIG. 36.—Area of circle.

MENSURATION—AREA AND VOLUME. 39

(b) The radius of a circle is $\frac{\text{diameter}}{2}$; hence if we write r for radius and d for diameter, $r^2 = \frac{d^2}{4}$. Also $\pi = 3\cdot 1416$ and $\frac{\pi}{4} = \cdot 7854$; hence $\pi r^2 = \frac{\pi}{4} \times d^2 = \cdot 7854\, d^2$. Thus in Fig. 37 the area of the circumscribing square A B C D, each side of which is 3 inches, is 9 sq. in.; the area of the circle is clearly less by the amount indicated by the shaded corners. If the area of these corners be estimated by counting the small squares (remembering that the area of each little square is ·01 sq. in.), they will be found to be nearly a quarter of the whole area, more exactly ·21; hence the area of the circle is *not* 3 × 3 but $3 \times 3 - \cdot 21\ (3 \times 3)$ approximately, or $3 \times 3 \times \cdot 7854 = 7\cdot 068$.

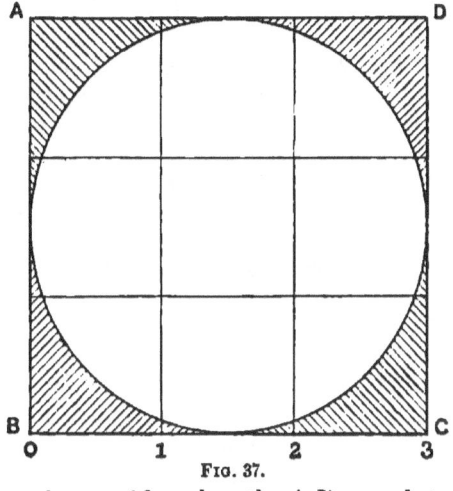

Fig. 37.

(c) Determine as in Experiment 13 a length A B equal to half the circumference of the circle = $\pi \times$ radius. Draw a rectangle A B E C, of which the sides C A and A B are half circumference and radius respectively; then area of rectangle equals area of circle = $\pi \times (\text{rad})^2$.

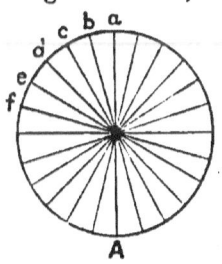
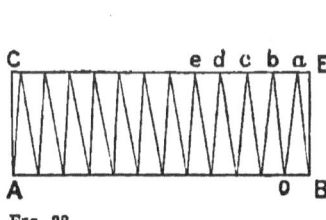

Fig. 38.

(d) Divide the circle into a number of small triangles $a\,b\,o,\ b\,c\,o,\ \ldots$, as shown in Fig. 38; arrange them so as to fill up the rectangle A B E C by placing all the triangles $a\,b\,o,\ c\,b\,o,\ \ldots$ which make up the semicircle $a\,b\,f\,A$, on the side E C of the rectangle, and a corresponding number on B A: these will be found to just fill up the rectangle.

MENSURATION—AREA AND VOLUME.

The area enclosed by any irregular boundary or figure, as shown at S (Fig. 39), can be estimated (approximately) when drawn upon squared paper by counting the squares wholly or partially enclosed by the boundary. A much better method is by the use of a planimeter. As these are easily obtainable and are usually fairly accurate, it is advisable to use them whenever possible.

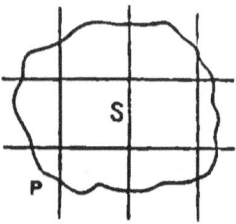

Fig. 39.—Area of an irregular figure.

Planimeter. — The area of any plane figure is most easily obtained by means of a planimeter. As will be seen on reference to Fig. 40, there are two limbs, each terminating in a needle-point, and resembling in some respects a pair of dividers. There is also a recording apparatus at W, consisting of a graduated wheel, with a vernier attachment and a graduated dial.

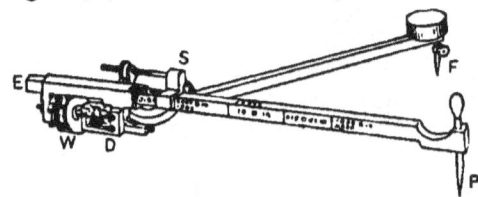

Fig. 40.—Planimeter.

When used to ascertain the area of any plane figure, the end F is fixed in some convenient position, so as to allow the point P to be easily carried round the profile of the figure whose area is to be determined. Whilst P is made to follow the boundary of the figure, the wheel W, and dial rotating, will at the end of the operation (when the point P starting from some point and tracing the boundary arrives at the starting-point) record the area.

Each rotation of the wheel W causes the dial D to move through $\frac{1}{10}$th of a revolution. As the wheel is divided into 100 equal parts and the vernier on the edge has a length equal to 9 of these divisions divided into 10 equal parts, it enables the reading to be read off to 3 places of decimals.

To employ the instrument to find the area of any irregular figure S (Fig. 39), the point F is fixed in such a manner that P can travel round the figure; then P is put into some point p; the readings of the wheel and dial are noted, and the boundary is carefully traced by P until p is again reached, and the readings of wheel and dial are now taken: the difference between these and the former readings will give the area required. Marks are to be found on the bar E P, so that the instrument can be adjusted by means of the screw S to give the area either in square inches or square centimetres.

MENSURATION—AREA AND VOLUME. 41

To test the accuracy of the instrument, and also to obtain the necessary freedom and proficiency in its use, it is advisable to draw a square or rectangle of any convenient dimensions, put the point P at an angular point, note the readings on the dial and the wheel, carefully trace the profile by the pointer P until the starting-point is again reached, again note the readings on dial and wheel; then the difference between this reading and the first will give the area of the figure. If P be taken round the figure again in the same direction, twice the area will be recorded. If it now be taken twice round in the opposite direction, the reading will be the same as at the commencement, if care has been exercised in travelling round the area.

Curved Surface of a Cylinder.

APPARATUS.—Wooden cylinder 3 or 4 units in diameter and 5 or 6 units in length; cone of the same dimensions; sheet of thin paper, and scale.

EXPERIMENT 23.—To find the lateral surface of a cylinder, and to prove that *lateral surface* $= 2\pi r \times l$, where r is radius of base and l is length of cylinder.

Take a strip of thin paper equal in length to the cylinder, wrap the paper tightly round the cylinder and mark by a pin or point of dividers two holes where the edges of the paper meet; unroll the paper and draw through the holes two parallel lines perpendicular to the edges of the paper, thus obtaining a rectangle, the base of which is the circumference of the cylinder and its altitude the length of the cylinder: ∴ *area* $= 2\pi r \times l$. To obtain the total surface, add on the areas of the two ends by Experiment 22.

Curved Surface of a Cone.

To verify that **the curved surface of a cone is half circumference of base × slant side** $= \pi r \times l$; or curved surface $= \pi r \times l$; area of base $= \pi r^2$ by Experiment 22; hence total surface $= \pi r l + \pi r^2 = \pi r (l + r)$.

If a cone be laid on a horizontal surface, as a table top, it can be made to roll, the vertex A remaining stationary.

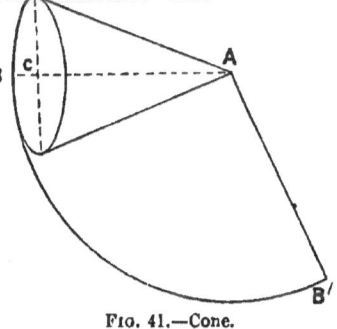

FIG. 41.—Cone.

Take any point B on the base of a cone (Fig. 41), and draw a straight line from B to the vertex A; on a sheet of squared paper draw a line equal in length to A B, and with A as centre. A B as radius, draw an arc of a circle; place the two lines so described together, and roll the cone until the point B meets the paper again at B'; join B' to A. Obtain the half-circumference of the base by Experiment 18; use the length so obtained as the base and l as the height of a rectangle; next obtain the area of the sector A B B'; show that the sector is equal in area to the rectangle.

APPARATUS. — Wooden cones 3 or 4 units diameter, 5 or 6 units in height; pencil compasses, scale, and squared paper.

EXPERIMENT 24. — Verify the rule for total surface of a cone (a) by calculation, (b) by rolling the cone on squared paper; (c) make a hollow cone of paper into which the solid cone will just fit, cut to shape, and verify as before.

Ellipse.

To verify the formula that the **area of an ellipse** = $\pi a b$, where a is the semi-major and b is the semi-minor axis of the ellipse.

Any of the well-known methods may be used to draw the curve. The following is a convenient method:—Draw two lines (Fig. 42) A O A' and B O B' at right angles to each other, and

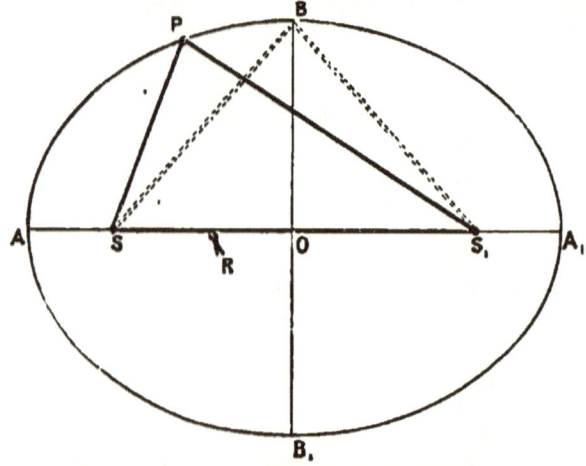

FIG. 42.—Ellipse.

make O A = O A' = semi-major axis, and O B = O B' = semi-minor axis. With centre B, radius O A, describe arcs of circles intersecting A A' at S and S'; the two points so obtained are called the foci of the ellipse. Put in as firmly as convenient

MENSURATION—AREA AND VOLUME.

three pins at S, B, and S' respectively; tie a piece of string tightly round the three pins; next take out the pin at B, insert a pencil, and, keeping the string taut, draw the curve. In any ellipse, $PS + PS'$ (or sum of the focal distances) has the same value for any point on the ellipse. From this property the above construction is derived. Using the same loop of string, and placing the pins further apart, the ellipse becomes elongated or flattened; as the pins are made to approach each other, the curve becomes more and more like a circle. When the pins are close together, the two foci coinciding, a circle is obtained.

If the ellipse be drawn upon squared paper, the area can be found approximately by counting the enclosed squares, as in the case of the circle. In Fig. 43 the two axes are 20 and 10 units respectively, the area is 157 square units; or if drawn upon cardboard, and the major and minor axes, A A' and B B', be drawn, then lines drawn parallel, and touching the curve, four rectangles are obtained, as shown in Fig. 43, each of

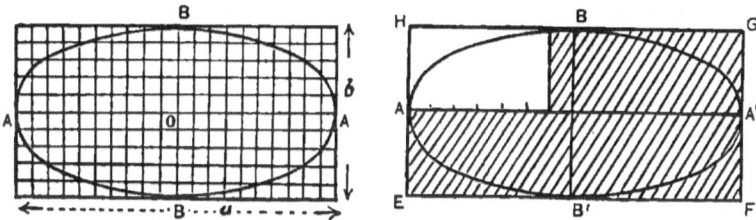

FIG. 43.—Area of an ellipse.

which is the rectangle on the semi-axis, making the shaded portion as shown equal to $3\frac{1}{7}$ of these rectangles, and cutting out the shaded portion, also cutting out an ellipse of the same size, the equality in the area can be obtained by weighing. This is explained on page 44.

EXPERIMENT 25.—Determine the area and mean height h of the hypothetical indicator diagram shown in Fig. 44. This can be obtained by drawing the diagram to scale, adding the 10 ordinates shown by dotted lines, using a long strip of paper and marking off at the end of the first the length of the second, and so on until the total length is obtained. This measured and divided by the number of ordinates (10) will approximately give the mean height h. The mean height h multiplied by the length of base A B will give the area. Or draw the curve on squared paper, and obtain the area by counting the enclosed squares; having thus obtained the area, compare mean height obtained by dividing the area by length of base A B with that previously obtained.

Areas by Weighing.

APPARATUS.—Sheet of cardboard, scissors, scale, and drawing instruments.

1. Cut out of the sheet of cardboard supplied to you a rectangle 4" × 2", or sides 4 cm. and 2 cm.; ascertain in each case, as accurately as you can, the weight in ounces or grams, and divide by 8. The number obtained will be the weight of a sq. in. or sq. cm.

2. Out of the same sheet cut out a circle 3" or 3 cm. diameter; verify by weighing that the area is (approximately) 7 sq. in. or 7 sq. cm.; or cut out one of the squares from the rectangle previously used, leaving 7 sq. units; put the rectangle into one scale pan and the circle into the other, and the two will be found to very nearly balance each other.

From the above sheet of cardboard in which the weight of a unit square is known, find by weighing the area of the following figures. Draw the figures on the cardboard, cut out carefully and weigh. The area of any irregular figure can be obtained in like manner.

3. An ellipse, major axis 10 cm., minor axis 6 cm.; and—
4. The given hypothetical indicator diagram (Fig. 44).

Measurement of Volume.

It is often necessary to read the scale division at the surface of the water in a narrow tube such as a burette; but the surface of the liquid, especially in a narrow vessel, is not flat, it is curved at every point where it touches the glass vessel. In the case of all the liquids with which we have to deal, except mercury, the surface is concave, as shown at A (Fig. 45). This will not affect the reading, if *the division of the scale is that which is opposite the lowest part of the curved surface. The eye should be placed on a level with the mark.*

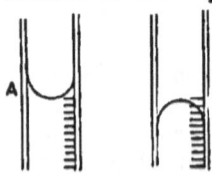

FIG. 45.—Capillary elevation and depression.

MENSURATION—AREA AND VOLUME. 45

In the case of mercury, the surface is convex as shown at B (Fig. 45), and similar precautions are necessary.

APPARATUS.—Sphere of any convenient size; burette graduated into cubic centimetres; stand; beaker (or tumbler) having its internal diameter slightly larger than the diameter of the sphere; right prisms of wood, brass, or other suitable material.

EXPERIMENT 26.—To find the volume of a sphere and to verify the formula. **Volume of sphere** $= \frac{4}{3}\pi r^3$.

For this purpose a burette, as shown in Fig. 46, may be used; this is usually graduated in tenths of a c.c. (cubic centimetre) in a downward direction, so that the number of c.c. run off is easily estimated. A small strip of gummed paper (such as a strip from a chemical label) may be used to make a mark on the side of the glass vessel, the distance of the mark from the base of the vessel being a little greater than the diameter of the sphere; so that when filled to the mark with water, the sphere is completely covered.

To obtain the volume of the sphere, having noted the height of the water in the burette, water is run into the glass vessel until the lowest part of the curved portion is coincident with the upper edge of the strip or mark. The

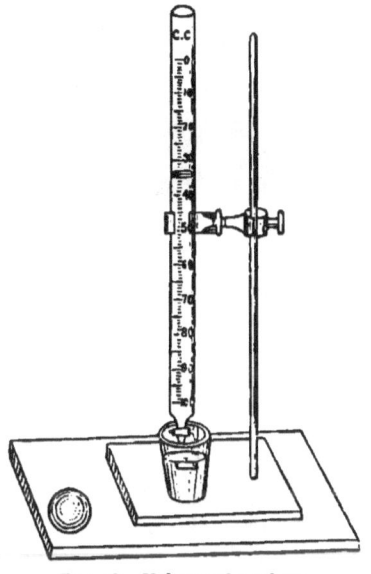

FIG. 46.—Volume of a sphere, or an irregular solid.

number of c.c. run off will give the volume of the vessel up to the mark; the water is emptied and the sphere inserted. It will be found that the number of c.c. of water required to fill to the same mark is less than in the previous case. The difference between this reading and the last gives the volume of the sphere in c.c. It will be found advisable to cover the surface of the sphere with oil, to prevent the water sticking to it. A pin fixed in the sphere may be used to lift it out or to hold it immersed.

With the outside calipers measure as accurately as possible the diameter of the sphere, and from the formula $\frac{4}{3}\pi r^3$ calculate the volume; compare the result with that obtained by experiment.

The volume may also be obtained by the use of a *measure graduated* in c.c.; the number of c.c. is observed, as shown at A (Fig. 47), then the sphere is carefully inserted. The number of c.c. displaced by the sphere is shown by the alteration in the height at B. In this, as in the preceding case, the lowest part of the curve is read off. The methods here described, and also the burette in Fig. 46, may be used to determine the volume of any irregular solid (F, Fig. 47), such as a pebble, a fragment of glass, etc., and in any case where it would be very difficult, if not impossible, to obtain the volume by direct measurement.

FIG. 47.

Volume of a Right Prism.—We have found that the area of a rectangle is the product of length × breadth; hence the volume of a solid rectangle is the product of *length × breadth × height*, or *area × height*.

Ascertain as accurately as you can the three dimensions of length, breadth, and height (by taking the mean of several readings), and obtain the volume by multiplying together the numbers so obtained; the numbers must all be expressed in the same units. Verify as shown in Experiment 26.

The volume of a cylinder = (*area of base*) × (*height*) = $\pi r^2 h$.

APPARATUS.—As in Experiment 26. In addition, a cylinder having its height and diameter equal to the diameter of the sphere.

EXPERIMENT 27.—As in the last exercise, obtain the number of c.c. of water required to fill the vessel when the solid is inserted, and subtract this from the volume of the vessel in c.c.; the difference gives the volume of the cylinder.

Verify by calculation, and tabulate—

Height of Cylinder ....
Diameter of Cylinder ....
Area of Base = πr^2 = ...
Volume = $\pi r^2 \times h$ = ...

MENSURATION—AREA AND VOLUME. 47

If the height and diameter of base of cylinder are each equal to the diameter of the sphere in the previous exercise, it will be found that the volume of the sphere is two-thirds that of the cylinder. Another method of proving this important result is as follows:—

Volume of a sphere = two-thirds the volume of the circumscribing cylinder.

APPARATUS.—A beaker, having its internal diameter equal to the diameter of the sphere.

EXPERIMENT 28.—Place a small strip of gummed paper on the beaker so that the distance from its upper edge to the bottom of the beaker is equal to the diameter of the sphere; divide the distance from the bottom of the beaker to the strip into three equal parts, and put a similar strip to the last at the first division a; fill up carefully to level a with water; insert the sphere, and the water will be found to rise to the upper level b.

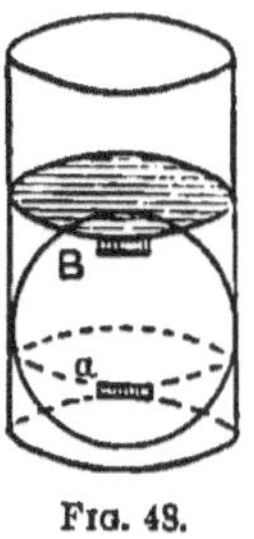

Measure as accurately as possible the diameter of the sphere by outside calipers; find by calculation the volume of the circumscribing cylinder. Two-thirds of this should equal the volume of the sphere.

FIG. 48.

The volume of a cone $= \frac{1}{3} \pi r^2 h = \frac{1}{3}$ (*volume of cylinder of same height and base*).

APPARATUS.—As in previous exercise. In addition, a cone of the same height and base as the cylinder.

EXPERIMENT 29.—Insert the cylinder, and note the volume of water required to fill the vessel up to the mark; remove the cylinder and insert the cone (it will be found that the volume of the cone is one-third that of the cylinder); note the volume of the cone in c.c., and verify the experiment by calculation, thus—

Base of Cone ....
Height of Cone ....
Area of Base ....
Volume..

Volume of a Cube $= a^3$.—A cube is a solid having six equal squares for its faces. A model can easily be made by cutting out of stiff paper or cardboard the six squares shown at S in Fig. 49, and bending along the lines shown. A solid is obtained as shown at S.

If the edge of the cube S be made of unit length (1 foot), then a larger cube of edge equal to 1 yard can be divided, each edge into

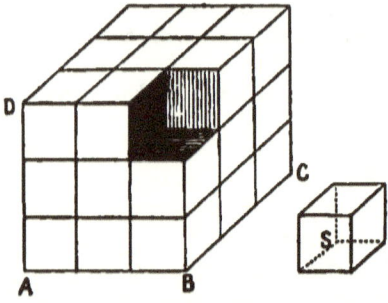

Fig. 49.—Volume of a cube.

3 equal parts as shown. The area of the base is seen to be 9 square feet, and as there are 9 perpendicular rows having 3 cubes in each row, there are altogether 27 cubes in all. Hence a cube having its edge equal to 3 units (inches, feet, or centimetres) has a volume of 27 cubic units—inches, feet, or centimetres.

If the edge of a cube be a units in length, the volume of the cube is $a \times a \times a = a^3$.

The Volume of a Pyramid $= \frac{1}{3}$ (*area of base*) $\times$ (*height*) $= \frac{1}{3} a h$.—If we imagine the centre o of the cube (Fig. 50) to be the common vertex of 6 pyramids, each of the 6 faces of the cube forming the base of a pyramid, the volume of each pyramid is $\frac{1}{6}$ the volume of the cube; hence the volume of each pyramid is $\frac{1}{6} a^3$. The height of each pyramid is $\frac{1}{2} a$; thus the volume of a pyramid can be written as $\frac{1}{3}$ (area of base) height. From this the well-known rule is easily remembered.

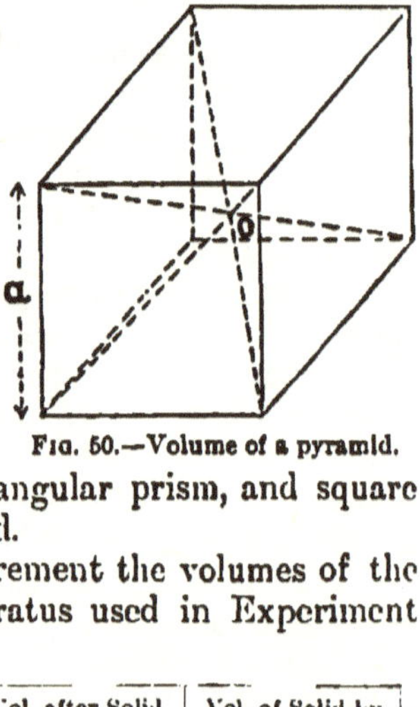

Fig. 50.—Volume of a pyramid.

APPARATUS.—As used in Experiment 26 or Experiment 28. The following solids: right prisms, cube, triangular prism, and square prism; also a cone and a square pyramid.

EXPERIMENT 30.—Obtain by measurement the volumes of the solids, and verify your results by apparatus used in Experiment 26 or 28. Tabulate as follows:—

Name of Solid.	Vol. by Measurement.	Vol. of Water before.	Vol. after Solid put in.	Vol. of Solid by Displacement.

MENSURATION—AREA AND VOLUME.

Volumes of Liquids.—One of the most convenient methods for measuring a volume of liquid is to use a graduated vessel, such as the burette described in Fig. 46; its use is sufficiently obvious. A **measuring flask** is shown in Fig. 51. As may be seen, it has a narrow neck with a mark round it. When filled with water to the mark, it contains exactly 1,000 c.c., or 1 litre, if the temperature be that indicated on the flask. As will be shown later, the volume of water is not constant, but expands when heated and contracts when cooled. Another instrument which may be used, called a **pipette**, is shown in Fig. 52. When filled to the mark at temperature 15° C., it contains 100 c.c. The pipette is also used to insert or withdraw a small quantity of liquid into or from a vessel.

FIG. 51.—Measuring flask.    FIG. 52. Pipette.

APPARATUS.—Balance and weights.

EXPERIMENT 31.—To test the relation in the metric system between the mass of a quantity of liquid and its volume. Counterpoise the weight of a beaker or tumbler with lead shot. From a graduated measure or burette run in 100 c.c. of water, add weights until equilibrium is restored; the added weights give the weight of 100 c.c. of water, from which the weight in grams of 1 c.c. or 1,000 c.c. can easily be calculated.

Measurement of Mass.—For the measurement of mass an ordinary *balance*, as shown in Fig. 53, or a spring balance, Fig. 59, is used. Thus, if two bodies are placed one in the pan P, the other in P_1, then if the force of gravity (or the earth-pull) is such that the two bodies balance each other, they contain equal quantities of matter. The balance consists of a lever very accurately made, having a hardened steel triangular knife-edge passing through the centre of the beam B, and resting on two flat steel plates on the top of the pillar p. When not in use the pans rest on the stand; the turning of the handle h raises the beam B, and allows

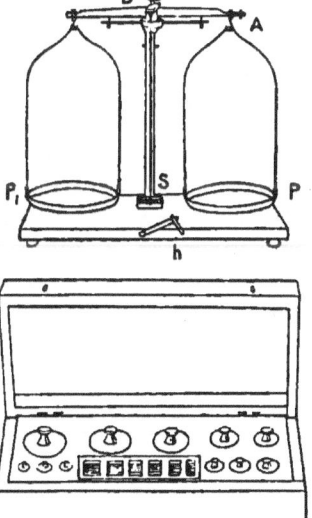

FIG. 53.—Balance and weights.

(1,004)

it to swing freely; a long pointer rigidly attached to the beam and moving in front of a graduated scale, S, enables the operator to judge when the two masses in the pans are equal; a small adjusting screw, a, enables the pointer to be set at zero when the pans are empty. A balance should be carefully examined; the manner in which the beam is supported by knife-edges at the centre, and the pans hung on knife-edges at the ends, should be noted.

In using a balance, the following rules should be observed:—

1. See that the pans are clean; dust may easily be removed by using a camel's-hair brush.

2. The pointer should be at the zero of the scale; any slight adjustment can easily be made by the adjusting screw a.

3. The body whose mass or weight is required must be put into the left-hand pan, and the weights put into the right, by means of the forceps; they must not be touched by the fingers. The body to be weighed and the weights used must be put in the pans only when the pans are resting on the stand.

4. The weights must not be placed anywhere except in the pans or in the box. For the fractional parts of a gram a small piece of cardboard is useful—this is ruled into squares, having the values of the smaller weights written on it; on this the weights can rest when not in use.

5. In using the balance, it is not necessary to wait until the pointer comes to rest; the weights in the pans are equal when the pointer swings to equal distances on each side of its mean or zero position. Having ascertained that the weights in the pan are sufficient, note the missing weights from the cardboard and the box, and check the calculation as these are replaced.

6. The balance must be handled with great care; if subjected to rough usage, it soon becomes useless for accurate work.

Comparison of English and French Weights.

APPARATUS.—Balance to weigh to ·01 gm. and to carry up to 100 gm.; also a common balance which will take weights up to 3 or 4 lbs.

EXPERIMENT 32.—Verify the latter part of Table I.; that is, complete the following:—

1 gm.	=	grains	=	ounces.
1 kilogram	=	ounces	=	pounds.
1 grain	=	grams.		
1 pound	=	grams.		

MENSURATION—AREA AND VOLUME.

Test the accuracy of the gram weights by putting the 50 gm. in one pan, the two 20 gm. and a 10 gm. in the other, etc.

Exercises.

1. Take a piece of copper or brass wire, measure its length, carefully form into a small-sized coil, determine its volume by the apparatus shown either in Fig. 46 or in Fig. 47, obtain the cross-section, and by Table II. the corresponding number on the wire-gauge.

2. Ascertain as accurately as you can the area of the internal cross-section and the bore of a glass tube, one end of which is sealed or closed by a cork, and which contains a known quantity in c.c. of water.

3. Measure the diameter of a piece of wire supplied to you by wrapping the wire round a rod (a lead-pencil will serve); each turn must lie closely against its neighbour. Measure the length l occupied by these turns; count the number of coils of the wire, which should be not less than 12. Divide l by the number of turns to get the diameter.

4. Using any convenient cylindrical tin or other vessel, to determine its internal volume in cubic inches to a certain scratch or edge of a strip of paper on its inner surface. Fill with water to the mark; find approximately the weight of a cubic foot of water in lbs.; also, as 1 cubic inch = 16·39 c.c., verify your weight in grams.

5. If a gallon measures 277·274 cubic inches, and a litre is represented by a cube whose edge is 3·937 inches, show that the number of pints in a litre is 1·76.

6. The major axis, A A', of an ellipse is 4"; the minor axis, B B', is 3": draw the curve, and prove that the area is approximately equal to that given by the formula, area = $\pi \times 2 \times 1\frac{1}{2}$.

Summary.

Circumference of a circle	= π (diameter) = $\pi d = 2\pi r$.
Area of a rectangle	= product of two adjacent edges a and $b = a \times b$.
Area of a square	= $a \times a = a^2$.
Area of a triangle	= $\frac{1}{2}$ (base) × (altitude).
Area of a parallelogram	= (base) × (altitude).
Area of a circle	= π × (radius squared) = $\pi r^2 = \frac{\pi}{4} d^2$.
Curved surface of a cylinder	= (circumference of base) × (length) = $2\pi r \times l$.
Curved surface of a cone	= $\frac{1}{2}$ (circumference of base) × (length of slant side) = $\pi r \times l$.
Area of surface of a sphere	= $4\pi r^2$.
Volume of a sphere	= $\frac{4}{3}\pi r^3 = \frac{2}{3}$ (circumscribing cylinder).
Volume of cylinder	= (area of base) × (height) = $\pi r^2 h$.
Volume of cone	= $\frac{1}{3}$ (area of base) × (altitude) = $\frac{1}{3}\pi r^2 \times h$.
Volume of a pyramid	= $\frac{1}{3}$ (area of base) × (altitude) = $\frac{1}{3}ah$.

A **balance** is used for the measurement of mass.

CHAPTER IV.

HYDROSTATICS.

Density and Relative Density, or Specific Gravity.—It is a common experience that equal volumes of different substances have different weights; thus, comparing four balls of equal size, of cork, iron, lead, and platinum respectively, although *the volumes* of the balls *are equal*, the iron ball is about thirty times the weight of the cork, the lead about one and a half times that of the iron, and the platinum nearly twice as much as the lead. From this we infer that the matter in the heavier bodies must be more closely packed together than in the lighter ones.

When two bodies of unequal size, but of the same material, are compared, the weights are directly proportional to the volumes; so that if we had two cubes of cast iron, the length of edge of one being twice that of the other, both the volume and the weight of the larger would be eight times that of the smaller.

If the three dimensions (length, breadth, and thickness) of a body are ascertained, either in centimetres or inches, the product will give the volume in cubic centimetres or in cubic inches; this multiplied by the weight of unit volume gives the weight. Conversely, if the weight and volume be known, the weight of unit volume can be obtained by dividing the former by the latter.

When subjected to great pressure, the volumes of the lighter bodies can be made very much less. Thus, a mass of cotton, which is a light and bulky material, under great pressure may have its density increased, and be made firm and compact as a piece of wood, and, like that material, it can be cut and planed. In a similar manner such materials as hay, straw, etc., are prepared for transport.

That property of matter by which equal volumes have unequal

weights is called *density*, and is used for comparing different substances with each other. The **density of a body is the mass of unit volume.**

In the metric system the unit volume is the **cubic centimetre,** and the **density** of any substance is **the number of grams in a cubic centimetre.** As *1 c.c. of water weighs 1 gram, the density of water is unity.*

In the English system the unit volume usually used is either the cubic inch or the cubic foot. A cubic foot of water weighs nearly 1,000 oz., or 62·5 lbs.

All bodies have the two properties of mass and volume; the mass is the product of the density and the volume, or

$$\text{density} \times \text{volume} = \text{mass}.$$
$$\therefore \text{density} = \frac{\text{mass}}{\text{volume}}.$$

The *practical measure* of the mass of a body is its *weight;* hence we may write the **density** as the **weight of unit volume,**

or
$$\text{density} = \frac{\text{weight}}{\text{volume}}.$$

It is often more convenient to refer to the relative density of a substance as compared with a standard. The standard adopted is water, on account of its being obtainable at any place in a pure state, its homogeneity, and its invariable density at a given temperature.

If the mass of any volume of a given substance be determined, and the mass of the same volume of water, the first divided by the second gives the **relative density,** or, as it is usually called, the **specific gravity** of the substance. A better term would be the **specific mass;** but as the intensity of gravity is the same for all kinds of matter, *the specific gravity* has the same value as *the specific mass.* Hence,

$$\begin{aligned}\textbf{Specific gravity} &= \frac{\text{mass of any volume of substance}}{\text{mass of equal volume of water}} \\ &= \frac{\text{weight of any volume of substance}}{\text{weight of equal volume of water}}.\end{aligned}$$

In the following table the density and specific gravity of a few substances are given :—

HYDROSTATICS.

TABLE III.

	Weight of a cubic foot in pounds.	Weight of a cubic inch in pounds.	Specific gravity.
Water.............................	62·5	·036	1·00
Lead...............................	711·6	·41	11·40
Platinum.......................	1343·9	·775	21·53
Wrought iron.................	485·6	·28	7·78
Cast iron........................	451	·26	7·23
Fir.................................	32	·018	·52
Mercury........................	848·75	·489	13·60
Copper...........................	552	·318	8·85
Oak................................	58	·034	·93
Cork..............................	15	·008	·24

APPARATUS.—Prisms or cubes of the following materials: cast iron, wrought iron, brass or copper, wood, and wax.

EXPERIMENT 33.—Weigh the prisms or cubes provided: their weights are different. Calculate the volume of each. Hence find the *weight of a cubic centimetre* and of a cubic inch; also find the specific gravity in each case (as the edges of the prisms or cubes may not be alike, the mean length, width, and depth must be obtained). Write these in columns headed as follows:—

Material.	Mean length. Mean width. Mean depth.	Volume in c.c. or in cubic inches.	Weight of prism or cube in grams or pounds.	Weight of 1 c.c. or 1 cubic inch.

1. Measure as accurately as you can the length, width, and thickness of the top of your work-table, and from Table III. find the weight in pounds or in grams.

EXAMPLE.—If the internal dimensions—length, width, and depth—of a cistern are 4½ feet, 3¼ feet, and 2⅓ feet respectively, what weight of water will it contain? If it contains 200 gallons, what is the height of the water?

$$\text{Volume} = \frac{9}{2} \times \frac{13}{4} \times \frac{7}{3} \text{ cubic feet.}$$
$$\text{Weight} = \frac{9}{2} \times \frac{13}{4} \times \frac{7}{3} \times 62\frac{1}{2} = \underline{2132\cdot8 \text{ lbs.}}$$
$$\text{Height} = 2\cdot18 \text{ feet.}$$

EXERCISES.

1. A cubic foot of copper weighs 560 lbs. It is rolled into a square bar 40 feet long. An exact cube is cut from the bar. What is its weight?

2. If 13 cubic feet of stone weigh one ton, find the weight of a cubic metre.

3. The mean diameter of a brass cylinder is 3·18 cm., its mean length 6·18 cm., its weight 413 gm. Find its volume and density.

$$\text{Diameter} = 3\cdot18; \text{ area} = \cdot7854 \times (3\cdot18)^2 = 7\cdot92 \text{ sq. cm.}$$
$$\text{Volume} = 7\cdot92 \times 6\cdot18 = 48\cdot96 \text{ c.c.}$$
$$\text{Density} = \frac{413}{48\cdot96} = 8\cdot43.$$

HYDROSTATICS.

4. Find the weight of a cast-iron ball 5 inches in diameter.
5. If a vessel holds a gallon of water, what weight of mercury will it contain?
6. If the internal dimensions of a tank are 5' × 4' × 2½', how many gallons of water will it hold?

Density of a Liquid.—This can be ascertained if the vessel is of such a shape that its volume, up to a given mark, can be determined by direct measurement. (A cylindrical vessel, such as that shown at A, Fig. 54, is convenient for the purpose.) Fill the vessel up to the mark with the liquid, and find its weight. This gives the weight of a known number of cubic centimetres of the liquid, and the weight of 1 c.c. can be found.

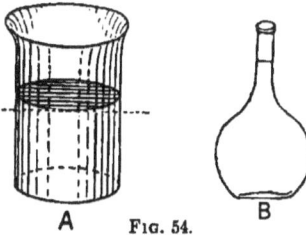

Fig. 54.

APPARATUS.—Any convenient cylindrical vessel; balance and weights; water.

EXPERIMENT 34.—Make a fine mark near the top of the vessel; find the mean internal diameter, the mean depth, and calculate its volume in c.c.

Counterpoise the vessel; then carefully fill it with water to the mark, and find its weight in grams.

$$\text{Weight of 1 c.c.} = \frac{\text{weight in grams}}{\text{volume in c.c.}}.$$

EXAMPLE.—If mean internal diameter = 3·18 cm., and height 6·18 cm.,

then area $= \frac{\pi}{4} \times (3 \cdot 18)^2$ = 7·92 sq. cm.
Volume = 7·92 × 6·18 = 48·96 c.c.
If weight when filled with water = 48·96 gm.,
then density $= \frac{48 \cdot 96}{48 \cdot 96} = 1.$

If convenient, instead of weighing the water, a known quantity (filling the vessel to the mark) can be run in from a burette, as shown in Fig. 46.

To obtain the weights of equal volumes of two liquids, either a flask or a beaker may be used. It is necessary for accurate comparison that the level of the liquid shall be coincident with a mark on the side or neck of the vessel; this may be either a scratch on the glass, or the upper edge of a strip of gummed paper attached to the glass. Owing to the curved form which the surface of the liquid assumes, it is impossible to prevent a small error in adjusting the level. If the cross-section of the vessel at the place where the mark is made is large, as shown at A (Fig. 54), then any small error in the height will mean a much greater error in the weight of the liquid than would be the case if the cross-section were narrow, as shown at B; hence the inaccuracy is diminished by the use of a narrow-necked vessel.

To determine the specific gravity of a liquid, two kinds of flasks—or specific-gravity bottles, as they are called—are used. One of these is shown at B (Fig. 54), having a volume as large as convenient, and terminating in a narrow neck, on which a mark is made.

When used to determine the specific gravity of a liquid, the flask is placed in the left-hand pan of the balance, and carefully counterpoised by small lead shot and tinfoil; the flask is next filled to the mark with pure water, again placed in the pan of the balance, and weights added until the balance is in equilibrium. The added weights denote the weight of the water. The flask is emptied and dried; when dry, it is filled to the mark with the liquid, and the weight obtained. Then

$$Specific\ gravity = \frac{\text{weight of liquid}}{\text{weight of water}}.$$

To avoid the trouble of weighing the water, the flask may be marked as shown in Fig. 55, the numbers indicating that at a temperature of 15° C. the volume of water which the flask contains when filled to the mark is 100 c.c. = 100 grams. So that if, as before, the flask is counterpoised, and the weight of the liquid in grams obtained, this, divided by 100, is the specific gravity required.

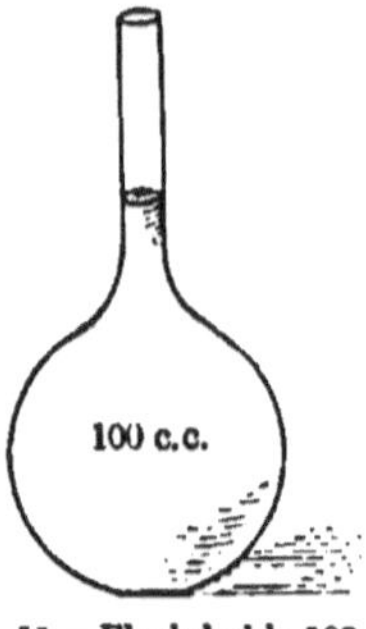

A more accurate and convenient method by which the specific gravity of a liquid can be obtained is by means of a **specific-gravity bottle**, shown in Fig. 56, which consists of a small bottle or flask with an accurately-fitting stopper through which a fine hole has been made. This allows the excess liquid to pass out when the stopper is inserted. When used to determine the specific gravity of a liquid, the bottle is carefully counterpoised and the weight obtained—(1) when filled with water, (2) when filled with the liquid.

To avoid the loss of time taken in filling and emptying the

FIG. 55.—Flask holds 100 c.c. of water.

HYDROSTATICS. 57

water and the subsequent drying, the weight of water which fills the bottle is often marked on it as shown; and if it contains, as shown in Fig. 56, exactly 100 c.c., then the weight of the liquid which fills the bottle, divided by 100, gives the specific gravity. For English measures, the number would indicate in grains the weight which fills the bottle.

The **specific-gravity bottle** may also be used to find the specific gravity of sand, small iron tacks, or any substance which can be obtained in fragments and is not soluble in water. This may be effected as follows:—

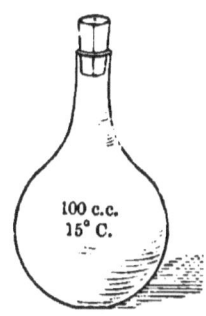

FIG. 56.—Specific-gravity bottle.

1. Weigh a small quantity of the substance the specific gravity of which is to be found. 2. Inserting this quantity in the bottle, it will displace a quantity of water equal to its own weight. The weight of the water displaced divided by the weight of the solid is the specific gravity required.

EXAMPLE.—The weight of a quantity of sand is 12 gm. when placed in a bottle, as shown in Fig. 56; and the bottle being filled up with water, the total weight, sand and water, is found to be 98 gm. Find the specific gravity of the sand.

Weight of a volume of water equal to that of the sand $= 112 - 98 = 14$ gm.

$$\therefore \text{specific gravity} = \frac{12}{14} = \cdot 857.$$

APPARATUS.—Balance and weights, 100 or 50 c.c. flask; turpentine, methylated spirit, distilled water, milk, hydrochloric acid.

EXPERIMENT 35.—Determine the specific gravity of the liquids supplied.

Tabulate your results thus:—

Liquid.	Weight when filled with water.	Weight when filled with liquid.	Specific gravity.	Weight of a c.c.	Weight of a cubic inch.
Methylated spirit...					
Hydrochloric acid..					
Milk..................					
Turpentine............					

EXERCISES.

1. An empty flask is counterpoised in a balance: when filled with water to a certain mark, the enclosed water is found to weigh 100 grains; when the water is replaced by turpentine, the weight is found to be 84 grains. Find the specific gravity of the turpentine.

2. A specific-gravity bottle when full of water is counterpoised by 983 grains in addition to the counterpoise of the empty bottle, and by 773 grains when full of alcohol. What is the specific gravity of the alcohol?

3. A bottle filled with water weighs 37·5 oz.; when filled with a liquid whose specific gravity is 0·96, it weighs 36·3 oz. Find the weight of the bottle.

4. The specific gravity of naphtha is 0·753. Find how many pints of naphtha weigh as much as a gallon of water.

5. A flask which holds 100 c.c. of water is filled with methylated spirit, and the weight of the liquid is found to be 79 grams. Find the specific gravity of the liquid.

Principle of Archimedes: *The upward thrust exerted on a body immersed in water is equal to the weight of water displaced.*—A body suspended by a fine string from one arm of a balance appears to lose a part of its weight when immersed in water, the loss in weight representing the weight of a volume of water equal to the volume of the body. Thus, if a body has the same density as water, *its apparent weight* in water is zero.

To prove the **principle of Archimedes** (Fig. 57), apparatus which consists of a hollow brass cylinder, into which a solid cylinder of the same material will just fit, may be used. The volume of water which the hollow vessel will hold is exactly equal to the volume of the solid cylinder. The hollow and solid cylinders are fastened to the end of a balance and carefully counterpoised; the solid cylinder is now detached, and fastened to a small hook provided for that purpose at the bottom of the hollow cylinder, and in such a position that the solid is totally immersed in water. The

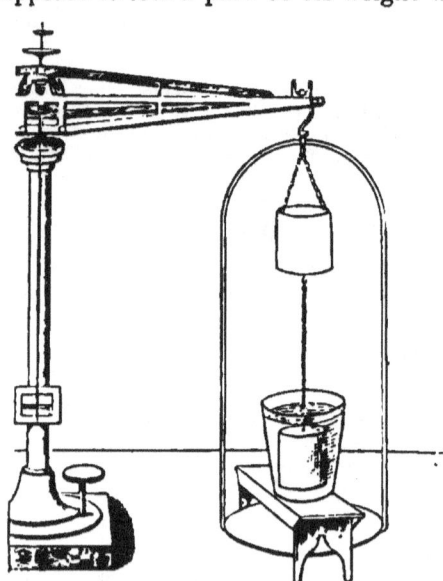

FIG. 57.—Verification of the principle of Archimedes.

HYDROSTATICS.

weight is now less, and the beam B will rise. By filling the hollow cylinder with water, equilibrium is restored when the cylinder is just full. Care should be taken that no air-bubbles adhere to the solid cylinder; also that it is totally immersed, with the upper edge a short distance below the surface of the water.

Specific Gravity of a Solid denser than Water. —The body is suspended from one end of a balance (Fig. 58) by means of a fine silk thread or silk fibre. Its *weight in air* and its apparent *weight in water* are determined. The *difference* between the weights so obtained is the weight of a volume of water equal to the volume of the solid.

FIG. 58.—Loss of weight in water.

Thus, if W_1 = weight in air, W_2 = weight in water,

$$\text{specific gravity} = \frac{W_1}{W_1 - W_2}.$$

Any air-bubbles which may adhere to the body when immersed in the water must be carefully removed by rubbing the body with a camel's-hair brush. A convenient method of determining the *specific gravity* or *relative density* is to use a spring balance, as shown in Fig. 59, the method being the same as before.

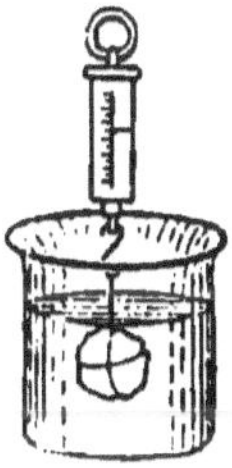

The graduated cylinder shown in Fig. 47 may be used: the height of the water in the cylinder before and after immersion is noted; thus the number of cubic centimetres of water displaced by the body is known.

APPARATUS.—Balance and weights, distilled water, beaker, brass or copper cylinders, one or more glass stoppers, etc.

FIG. 59. Loss of weight in water.

EXPERIMENT 36.—Determine the specific gravity of the solids supplied to you. Tabulate your results as follows:

Material.	Weight of solid in air.	Weight of solid in water.	Weight of an equal volume of water.	Specific gravity.
Brass cylinder.......				
Copper cylinder.....				
Glass stopper.........				
....				
....				
....				

HYDROSTATICS.

Specific Gravity of a Body less dense than Water.—A solid less dense than water, such as a piece of wood, will float; hence its apparent weight in water is zero. To determine the specific gravity, a *sinker* (which is a piece of a dense material, such as lead) is tied to the solid: it should be large enough to secure complete immersion of the solid.

To determine the specific gravity of the solid, (1) find the weight of water displaced by the sinker; (2) find the weight of water displaced by the sinker and the solid. Hence find the weight of water displaced by the solid.

Thus, if weight of solid in air $= W_1$
Weight of solid and sinker in water $= W_2$
Weight of sinker in water $= W_3$

$$\text{Specific gravity} = \frac{W_1}{W_1 - W_2 + W_3} = \frac{W_1}{W_1 - (W_2 - W_3)}.$$

APPARATUS.—Balance and weights, distilled water, beaker, fine thread for suspension, beeswax, small wood blocks, a wooden pill-box, small shot, etc.

EXPERIMENT 37.—Find the specific gravity of the materials supplied. Enter your results in the following table:—

Material.	Weight of solid in air.	Weight of sinker alone in water.	Weight of solid and sinker in water.	Specific gravity.
Beeswax........				
Wood				
....				
....				

(A sinker is easily made from a piece of sheet lead, through the centre of which a piece of wire, bent to form a hook, is inserted.)

Load the pill-box with small shot. Obtain the total weight of box and shot. Float in water, in measuring flask (Fig. 47), and note that c.c. of water displaced is equal to the weight in grams.

APPARATUS.—Balance and weights, beaker, milk, turpentine, methylated spirit, beeswax, glass stoppers.

EXPERIMENT 38.—Determine the specific gravity of the liquids supplied. This is done by finding the weight of any volume of the liquid and the weight of an equal volume of water, taking

HYDROSTATICS. 61

any solid, such as a glass stopper or the brass cylinder in Experiment 36. Find the weight of water and the weight of the liquid displaced by the solid. From these weights the specific gravity can be found.

EXAMPLE.—Weight of cylinder in air   = 413 gm.
Weight of cylinder in water   = 359·2 gm.
Weight of cylinder in glycerine = 345·8.
Specific gravity of liquid = $\dfrac{\text{weight of liquid displaced}}{\text{weight of water displaced}} = \dfrac{67\cdot2}{53\cdot8} = 1\cdot25$.

EXAMPLE.—A cylinder when weighed in water is found to displace 26·7 gm. of water, and in turpentine to displace 23·2 gm.
$$\therefore \text{specific gravity of turpentine} = \dfrac{23\cdot2}{26\cdot7} = \underline{·868}.$$

EXERCISES.

1. A piece of brass which weighs one kilogram, when weighed in water is found to weigh 879·5 gm. Find the specific gravity of the brass.
Weight in air 1000 gm.
Weight of a volume of water equal to that displaced by the kilogram = 1000 − 879·5.
$$\text{Specific gravity} = \dfrac{1000}{1000 - 879\cdot5} = \dfrac{1000}{120\cdot5} = \underline{8\cdot3}.$$

2. The weight of a body *in vacuo* is 3 lbs.; a piece of metal whose specific gravity is 10 is fastened to it. If the piece of metal weighs 1 lb., the compound piece will just sink in water. Find the specific gravity of the body.

3. A piece of round wire is 30 cm. long, its weight in air is 24·227 gm., and in water 21·121 gm. Find its diameter.
The volume = 24·227 − 21·121 = 3·106 c.c., since 1 c.c. = 1 gm.
But if r = radius, then volume = $\pi r^2 \times 30 = 3\cdot106$.
$\therefore r = ·1815$ cm., and diameter = $2 \times ·1815$.
= ·363 cm.
= $\underline{3\cdot63 \text{ mm.}}$

4. A piece of copper wire 4 metres long weighs 17·2 gm. in air, and 15·2 gm. in water. Find the diameter of the wire.

5. A glass rod which weighs 35 gm. in air weighs 21 in water. What will be its apparent weight in alcohol of specific gravity 0·9?
Let W denote the weight.
$$\text{Specific gravity} = \dfrac{35}{35 - 21} = 2\cdot5.$$
$$\therefore W = 35 - \left(\dfrac{35}{2\cdot5} \times ·9\right)$$
$$= \underline{22\cdot4 \text{ grams.}}$$

6. A piece of wood weighs 150 gm. in air, the sinker weighs 75 gm. in water; the two together in water weigh 45 gm.
$$\text{Specific gravity} = \dfrac{150}{150 - 45 + 75} = \dfrac{150}{180} = \dfrac{5}{6} = ·83. \ldots$$

7. Find what weight of lead attached to 20 lbs. of cork, specific gravity ·24, would be just sufficient to sink it in water.

Specific Gravity of a Liquid by Means of a Balance.—The specific gravity of a liquid can be found by means of the balance, if we take any solid which is not acted upon chemically by either water or the liquid, and which will sink both in the water and in the liquid. (1) Weight of solid in air, call this W; (2) weight in water, W_1; (3) weight in liquid, W_2.

$W - W_1$ = weight of an equal volume of water.
$W - W_2$ = weight of an equal volume of liquid.

$$\text{Specific gravity} = \frac{W - W_2}{W - W_1}.$$

8. A glass ball weighs 1,000 grains in air; it weighs 630 grains in water, and 650 grains in wine. Find the specific gravity of the wine.

$$\text{Ans. } \frac{1000 - 650}{1000 - 630} = \frac{350}{370} = \cdot 946.$$

Jolly's Balance.—In this balance a spiral spring S carries two light scale pans P and P'. On the face of the pillar A a scale etched on mirror-glass and graduated in millimetres is placed. The index is a small white bead b. The lower pan is always immersed in a beaker or tumbler containing water. It will be found convenient to attach (to the wire carrying the lower scale pan) a fine, bent wire w, and to see that the wire in each case touches the surface of the water before a reading is taken. The reading on the scale is now taken (both pans empty).

The body whose specific gravity is required is put into the pan P, the platform D is lowered by means of the screw S_1 until the pan P' is a short distance from the base of the beaker, and the scale reading is observed. The body is now placed in the lower pan, and the reading again noted.

From the three readings the specific gravity is obtained.

Thus, let a denote the reading when both pans are empty; let b denote the reading when the body is in the top pan; and let c denote the reading when the body is in the bottom pan.

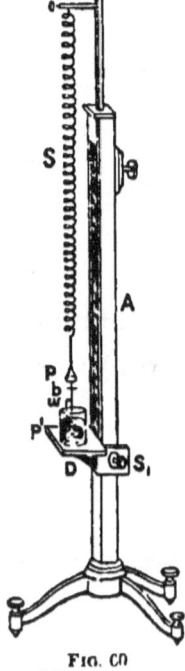

Fig. 60.
Jolly's balance.

Then specific gravity $= \dfrac{b - a}{b - c}$.

HYDROSTATICS.

U-Tube.—A U-tube, such as that shown in Fig. 61, is easily made by bending a piece of glass tubing, about $\frac{1}{4}''$ or $\frac{3}{8}''$ bore and about 36" long, over a bat's-wing flame. The tube may be fastened to a wood support by thin sheet brass clips, bent so as to embrace the tube, and fastened by screws to the wood. Two scales, which may be graduated either in inches or millimetres, are attached as shown. Water (which may be coloured by adding a little aniline dye) is poured in, the height in the two limbs being the same.

Manometer.—Attach one end A by an india-rubber tube to a gas tap. Turn on the gas; the column in A will be depressed, that in B will be raised: the difference in the height above any horizontal line, such as C D, will give the pressure of the gas (in inches or millimetres of water). A U-tube used in this manner is called a *manometer*. If, instead of water, a heavier liquid, such as mercury, be used, the difference between the heights is less than before. but is greater if a lighter liquid, such as turpentine, be used.

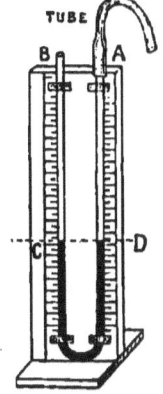

Fig. 61.
U-tube used as a manometer.

Apparatus.—U-tube; water, mercury, turpentine, methylated spirit, glycerine.

Experiment 39.—Fill the U-tube about half full of water; attach one end by an india-rubber tube to gas supply; note the readings on the scales, and deduce the pressure of the gas above the atmosphere.

The heights may be measured from any horizontal line, such as C D; or draw a line across at the level of the lower column, and thus obtain the difference in the heights.

Tabulate as follows:—

Liquid.	Height in open limb.	Height in gas limb.	Difference.	Pressure of gas above atmosphere.	Pressure of gas per sq. in.
Water........	20·7 cm.	13·5 cm.	7·2 cm.	72 mm. of water	
Methylated spirit......	20·7 cm.	12·5 cm.	8·2 cm.	82 mm.	
Turpentine..	....	....	....	....	
Glycerine....	....	....	....	....	

Specific Gravity by Means of U-Tube.

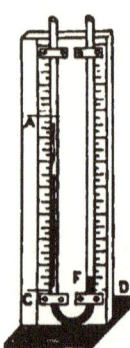

Fig. 62.—Specific gravity by U-tube.

—When two liquids which do not mix are put into the vertical limbs of a U-tube, the heights above any horizontal line, C D, are found to be different; by means of the scales attached, *the heights* of the two columns from the surface of separation of the two liquids can be determined. These will be found to be *inversely proportional to their densities.*

Thus, if a column of water C A, 50 cm. high, is balanced by a column of mercury D F, 3·67 cm. high (Fig. 62), the density of the mercury $= \dfrac{50}{3 \cdot 67} = 13 \cdot 6$ (∴ 13·6 volumes of water balance 1 volume of mercury).

APPARATUS.—U-tube with scales attached; water, mercury, methylated spirit, glycerine, turpentine.

EXPERIMENT 40.—Pour mercury into the tube until there are about 4 cm. in each limb, and pour a small quantity of water into limb A. Measure the difference in the heights of the water and the mercury above the fixed line C D. Increase the amount of water in A and again measure the heights. Proceed in a similar manner when more water is put in at A. Tabulate thus:—

Liquid.	Head or height of water.	Head of liquid.	Specific gravity of liquid.
Mercury	73·2	5·4	13·56
,, ...			
,, ...			
Methylated spirit.........			
Turpentine.........			
Glycerine.........			

EXERCISE.

1. Find the specific gravity of glycerine, having given that the glycerine barometer reads 319 inches when the mercurial barometer reads 29·66 inches, the specific gravity of mercury being 13·568.

HYDROSTATICS. 65

Specific Gravity of a Liquid by the inverted U-Tube (*Hare's Apparatus*).—Two long glass tubes are supported in a vertical position by means of clips and a suitable wooden stand, as shown in Fig. 63. Pieces of india-rubber tubing connect two open ends of a T-piece A to the vertical tubes; to the remaining end of the T-piece a piece of rubber tubing (which can be closed by a clip) is attached. The tubes are placed in front of two scales divided into millimetres; the open ends of the tubes dip into two small beakers (or bottles), one of which contains distilled water, the other the liquid whose specific gravity is to be found.

By suction at the end of the piece of india-rubber tubing the liquids can be drawn up into the tubes, until one of them is not far from the top. (When liquids are used which wet the tubes, they should be drawn up and down a few times before readings are taken.) The beakers are then adjusted in height so that the surfaces of the liquids are coincident with the zero of the scales; hence the scale readings give the height of the columns above the surfaces in the beakers. *The heights* so obtained are *inversely* as *the densities of the liquids;* thus the *ratio of the densities is the reciprocal of the ratio of the heights.*

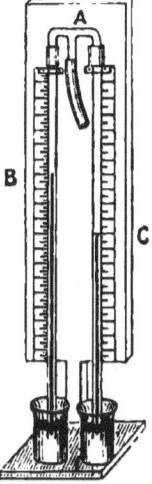

FIG. 63.—Specific gravity by Hare's method.

APPARATUS.—Hare's apparatus; alcohol, turpentine, milk, etc.

EXPERIMENT 41.—Determine the specific gravity of the liquids supplied you.

EXERCISE.

Height of alcohol in limb B = 36·2 cm.
Height of water in limb C = 28·9 cm.

Specific gravity of alcohol $= \dfrac{28\cdot 9}{36\cdot 2} = \cdot 8 \ldots$

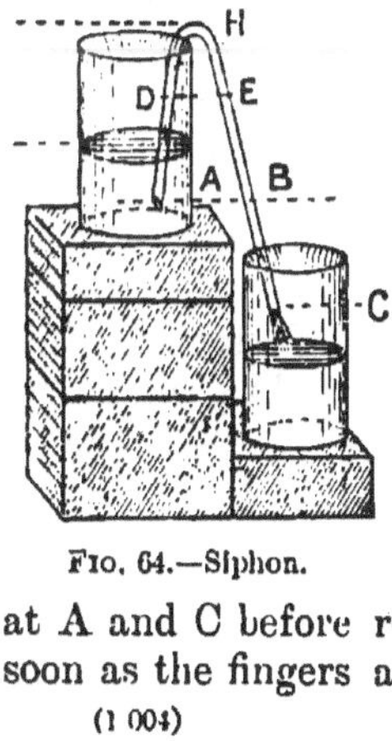

APPARATUS.—Glass U-tube; water in two beakers, and some blocks of wood. Arrange the wooden blocks so that the surface of the water in A is higher than that in C. (Fig. 64.)

EXPERIMENT 42.—Fill the U-tube with water; close the open ends with the fingers, and invert, putting the ends into the water at A and C before removing the fingers: it will be found that as soon as the fingers are removed water commences to flow from A

FIG. 64.—Siphon.

to C; the flow ceases when the surfaces in the two beakers are at the same height. A little aniline dye or a few drops of ink put into the beaker A will enable the flow to be seen clearly. If now the surface in A is lowered by removing some of the blocks, the reverse action takes place—the water passing from C to A, and, as before, the flow is maintained until the heights of the surface are equal. If the end at A be raised slightly above the surface, the water instantly leaves the tube.

The Siphon in its simplest form consists of a bent tube similar to the above, but has usually one leg longer than the other; this is not, however, important, the direction of the flow depending only on the levels of the two ends. Thus, if the end at C is lower than that at A (Fig. 62), the liquid may be transferred from A to C—that is, from a higher to a lower level.

To start the siphon the tube is filled and inverted. But it is not necessary to have the end at C below the surface of the liquid; all that is required is to put the end of the tube below the surface in A before removing the finger, and to keep the end at C below the level of A; when this is done the flow commences as soon as the finger is removed from the end A, and may be continued until the beaker is emptied.

If any horizontal line D E be drawn, then as the pressure on the water in the beaker A is equal to that of the atmosphere, the pressure at D is less than the pressure at A by the weight of a column of water between D and the level of the water in A. The pressure at E, which is at the same level as D, is less than the pressure of the atmosphere by the weight of a column of liquid from E to the level of the water in C. The pressure on the end at A is greater than the pressure at C by the difference in the lengths of A H and H C, so that the water will flow from A, the place of high pressure, to C, that of lower pressure. It is evident that the highest point, H, of the tube should not exceed the height of the water barometer. Instead of having to fill and invert a siphon, a simple arrangement is to use two taps, one in each branch of the siphon, so that when closed a column of liquid is enclosed; when placed in position and the taps opened, the flow commences, and is maintained as long as they remain open.

EXERCISE.

Write out a few useful applications of the siphon; also state what modifications would enable a siphon to act without having to fill and invert before being used.

HYDROSTATICS.

Head of a Liquid.—If an orifice be made in the side of a vessel containing water, it is well known that the water will flow from it; if the orifice be closed by inserting a cork, then by increasing the height of water in the vessel it will be found that at a certain height the cork is forced out, showing the increase in the pressure due to the increased height (or head) of the water.

The pressure at any level is due to the water above that level pressing upon the water below; hence if the sides of the containing vessel were vertical, the pressure at any level, and therefore the pressure on the base, would be the whole weight of water above.

A very convenient method of showing that the pressure increases with the depth is shown in Fig. 65. A glass or metal funnel, of any convenient size, is placed in a tall beaker or cylinder the diameter of which is large enough to allow the funnel to move freely in it. If now the end of the funnel be connected by means of an india-rubber tube with the U-tube shown in Fig. 61, then by means of a scale placed alongside the depth of the surface of the water in the funnel F can be estimated, and the corresponding pressure is given by the indication on the scale of the U-tube.

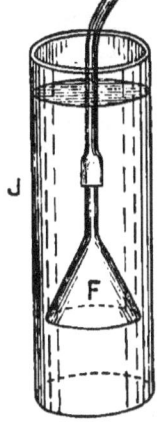

Fig. 65.—Apparatus to show pressure due to depth.

APPARATUS.—U-tube; funnel; india-rubber tubing; tall glass beaker or cylinder.

EXPERIMENT 43.—Fill the beaker nearly full of water; fasten one end of the rubber tube to the funnel, and connect the other to the U-tube; show, by lowering the inverted funnel, that the *pressure increases with the depth*.

Another and perhaps a better method than that described is to fasten, by means of a piece of fine thread or wire, a small india-rubber balloon to one end of a straight glass tube, and to connect

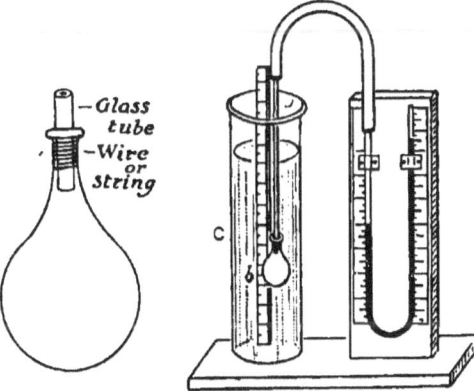

Fig. 66.—Apparatus to show pressure due to depth.

68 HYDROSTATICS.

the other end by a piece of india-rubber tubing to one branch of the U-tube. The balloon *b* (Fig. 66) is filled with air (the enclosed air can be prevented from escaping by using a small clip on the flexible tube) and attached to one end of the U-tube. When placed in a tall jar or cylinder of water C (Fig. 66), the pressures on the enclosed air in *b*, due to the pressures of the water at various depths, will cause the column of water in the U-tube to descend on one side and to rise on the other. By means of a scale placed in the jar C, and the scales attached to the U-tube, the depth of *b* and the corresponding pressure at that depth can be ascertained.

The results obtained should be tabulated as follows :—

Depth of *b* in Inches.	Reading on scale in cylinder C.	Readings on scales of U-tube.	Pressures at the various depths.

Hydrometers.—This is the name given to instruments which are chiefly used to determine the specific gravity of liquids. This is effected either by observing the depths to which the hydrometer sinks when placed in the liquid, or by means of weights which can be placed on a suitable pan, and which will sink the instrument to a standard mark on the stem.

APPARATUS.—Test-tube, small shot or mercury, glass tubing.

EXPERIMENT 44.—Load the test-tube with small shot or mercury until it sinks to a convenient point in water, mark the point by scratching with a file, add salt to the water, and ascertain the difference in the density of the water shown by the rise of the simple hydrometer; immerse in other liquids of known specific gravity, and mark the points. Blow a bulb on one end of a piece of glass tubing, allow it to cool, load as before and repeat the experiments.

The common hydrometer (Fig. 67) consists of a graduated glass tube, having at one end two bulbs A and B—one containing air,

HYDROSTATICS.

and rendering the instrument specifically lighter than water; the other containing sufficient mercury to make the instrument float upright when placed in water.

The **specific gravity of any liquid** into which the instrument is placed **is indicated by the depth to which it sinks,** and this can be read off by the scale, which is often divided into 100 parts and marked on or enclosed in the stem.

It may be used to determine the specific gravity of liquids either heavier or lighter than water.

In the case of liquids heavier than water, the amount of mercury in the lower bulb is so adjusted that the *highest* division of the scale is coincident with the surface of the water. The divisions on the scale are not equal, those towards the bottom being shorter than those above. The scale divisions are easily obtained by immersing the instrument in liquids of known specific gravities and marking the points.

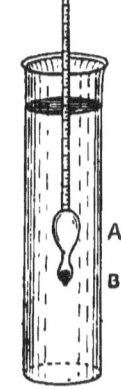

FIG. 67.
Hydrometer.

When a hydrometer is used to determine the specific gravity of liquids lighter than water, the adjustment is so made that the *lowest* division on the scale is level with the surface of the water; as before, the scale divisions are not equal in length, the distance between them becoming less as the top of the scale is reached.

Thus in each case the beginning of the scale is that division which is coincident with the surface when the instrument is immersed in water.

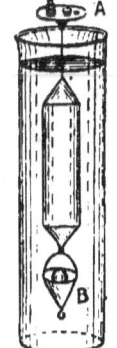

FIG. 68.
Nicholson's hydrometer.

Nicholson's hydrometer consists of a hollow brass cylinder, carrying at one end, by means of a wire stem, a scale pan A (Fig. 68), and at the other, by a stirrup, a pan B. When placed in a liquid the instrument floats, with the axis of the cylinder vertical, the pan B below and the pan A above the surface of the liquid, as shown. On the wire which carries the pan A a mark or standard point is made; the weight in the upper pan required to sink it to this point is called the standard weight. To find the weight of any small body, it is first placed in the upper pan; weights are added until the instrument sinks to the standard point; the difference between this and the standard weight is the weight of the body. If the body be now placed in the lower pan, on account of the upward pressure of the water more weight is required to sink the instrument to the standard point; the difference between this weight and the last is the weight of water displaced by the body. $\text{Specific gravity} = \dfrac{\text{weight of body}}{\text{weight of an equal volume of water}}$.

EXAMPLE.—The standard weight being 1,200 grains, a body is placed in the upper pan, and it is found that 200 grains must be added to sink the instrument to the standard point. The body is now placed in the lower pan, and 450 grains must be placed in the upper pan to sink the instrument to the standard point. What is the specific gravity of the body?

Weight of body = 1,200 − 200 = 1,000 grains.
Body in lower pan = 450 − 200 = 250 grains.

This is, by the principle of Archimedes, equal to the weight of water displaced by the body.
Thus the weight of the body is 1,000 grains;
The weight of an equal volume of water is 250 grains.

$$\therefore \text{specific gravity} = \dfrac{1000}{250} = 4.$$

SUMMARY.

Density.—Equal volumes of the same substance have equal masses or weights; equal volumes of different substances have different masses or weights.

The volume of a body in cubic centimetres is also the volume of water in cubic centimetres displaced by it when immersed.

Specific gravity of a substance $= \dfrac{\text{weight of substance}}{\text{weight of water displaced}}$.

Specific gravity of a liquid $= \dfrac{\text{weight of any volume of liquid}}{\text{weight of an equal volume of water}}$.

Specific gravity of a liquid by U-tube $= \dfrac{\text{height of water column}}{\text{height of liquid column}}$.

Specific Gravity by U-Tube.—The "*heights*" of the two columns are *inversely proportional to their densities.*

Specific Gravity by inverted U-Tube (*Hare's apparatus*).—The specific gravities of the two liquids are *inversely proportional to the* "*heights.*"

Principle of Archimedes.—When a body is wholly or partially immersed in a liquid, the resultant upward thrust is equal to the weight of the liquid displaced. The solid floats when its weight is less than the upward thrust due to the liquid displaced, and sinks when greater.

Hydrometer readings on graduated stem, when placed in a liquid, give specific gravity.

Nicholson's Hydrometer.—At one end a scale pan, at the other a bucket or pan; is always sunk to a fixed point, which is marked on the stem.

CHAPTER V.

PROPERTIES OF AIR—BAROMETERS—BOYLE'S LAW.

Properties of Air.

AIR is a mixture of several gases, but it is found to act in nearly all physical respects as an elementary, permanent gas. That air is a form of matter may be shown by inserting a tumbler into water with the open end downwards. It will be noticed that the water does not fill the tumbler, showing that something occupies the space within the tumbler; water and air cannot occupy the same space at the same time. The *elasticity* of air is also shown by the rise of the water and the diminished volume of the enclosed air as the tumbler is pressed still further into the water. That air has *weight* can be shown by weighing a vessel from which the air has been exhausted by means of an air-pump. By means of a tap, T (Fig. 69), the air can be admitted, and the vessel again weighed; the difference is the weight of air filling the vessel. If now the weight of an equal volume of water be obtained, the density of *air* can be found.

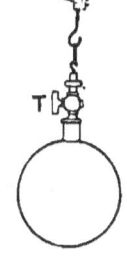

FIG. 69. Apparatus to show air has weight.

EXERCISE.

An empty flask weighs 120 grams; when full of air it weighs 121·3 grams, and when filled with water 1,120 grams. Find the density of air.
  Weight of enclosed air   = 121·3 − 120 = 1·3 grams.
  Weight of equal volume of water = 1,120 − 120 = 1000 grams.

  Specific gravity of air $= \dfrac{1·3}{1000} = ·0013.$

But 1 c.c. of water weighs 1 gram, ∴ density of air = ·0013; hence 1 litre of air weighs 1·3 grams.

It will be noticed that the apparent weight in each case is less than the true weight by the weight of air displaced by the flask; but as this is the same at both the observations, it does not affect the difference of the weights. On account of the compressibility of air, the density of air—that is, weight of unit volume—depends upon both the *pressure* and the *temperature*.

APPARATUS.—Glass flask, and glass and rubber tubing.

EXPERIMENT 45.—Fit an india-rubber cork and a short length of glass tubing into the neck of a glass flask containing a small quantity of water; to the glass tube a piece of india-rubber tubing which can be closed by a pinch-cock is attached. If by the application of heat (by a Bunsen burner) steam be formed and allowed to issue freely, all the air can be driven out of the flask. The flask is allowed to cool, and its weight is ascertained; then the increase in weight when air is allowed to enter can easily be determined.

Barometer.

APPARATUS.—Glass tube about 36" long, about $\frac{3}{8}$" bore; mercury; evaporating-dish.

EXPERIMENT 46.—Nearly fill the tube with clean mercury, leaving about half an inch empty; close the open end carefully with the finger, and invert, allowing the bubble of air to pass to the top; repeat this until all the small bubbles of air clinging to the glass are swept out. Fill up the tube quite full of clean mercury, and putting the finger on the open end, invert in a small evaporating-dish containing mercury, taking care that the open end is below the surface before removing the finger. It will be found that the column of mercury will descend a short distance, until the height of the column above the surface in A (Fig. 70) is about 30 inches, or 760 millimetres. This column is balanced by the downward pressure of the atmosphere, which, acting on the mercury in the dish and being transmitted equally in all directions, acts upwards on the column in the tube. When the column B ceases to descend, the mercury column and the atmosphere just balance. The space above the mercury contains no air and only a slight trace of mercury vapour, and is therefore a nearly perfect vacuum.

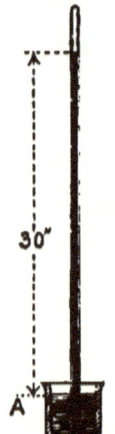

FIG. 70. Height of mercury in barometer.

The Barometer is an instrument which records (or shows) the pressure of the atmosphere, and in its simplest form consists of the Torricellian tube just described, in which for any increase of pressure the mercury rises, and falls when the pressure is diminished; *thus the height of the barometer measures the atmospheric*

PROPERTIES OF AIR—BAROMETERS—BOYLE'S LAW.

pressure. Any other liquid instead of mercury could be used, but mercury is the heaviest known liquid, being 13·6 times as heavy as water. A column of water to balance the mercury column would have to be 13·6 times as high. If water were used, the height would be $\dfrac{13\cdot 6 \times 29}{12} = 33$ feet (approximately). Other advantages of mercury, in addition to convenience in height, are that it remains liquid until it is cooled to − 40° C. (when it becomes solid); it does not wet the sides of the tube, and therefore moves easily in it; its density has been carefully determined for a large range of temperature, and corrections for temperature can thus be made.

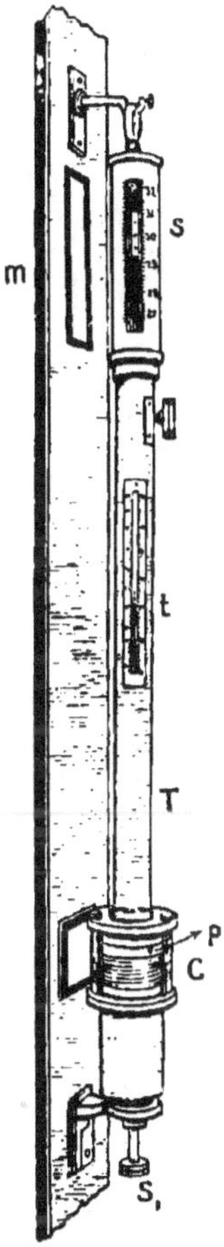

EXERCISE.

If glycerine (S.G. 1·26) be used in a straight tube, show that the height will be about 27 feet when the water barometer is at a height of 34 feet.

Water, although more sensitive than mercury or glycerine, is unsuitable as a liquid for a barometer. Water is found to evaporate into the vacant space, and air absorbed by the water also interferes with the action. These objections do not apply to glycerine. A glycerine barometer is more than ten times as sensitive as a mercurial barometer.

Fortin's Barometer.—In the barometer just described, as the mercury in the tube rises, the level of the mercury in the cistern falls; and, conversely, when the mercury falls in the tube, it rises in the cistern. If the area of the tube be small in comparison with that of the cistern, there will be only a slight alteration in the level of the mercury in the cistern, due to the rise or fall in the tube, and this is often neglected; but for accurate work it must be taken into account. In Fortin's barometer (shown in Fig. 71), the error is corrected by raising or lowering the surface of the mercury in the cistern. The tube T and the cistern C are enclosed in a metal case for protection; at S a graduated scale and a vernier enables the reading to be taken to $\frac{1}{500}$ of an inch, a thermometer t indicating the temperature at which the reading is taken. By means of the screw S_1, the surface of the mercury in the cistern

FIG. 71.—Fortin's barometer.

can be adjusted so that it touches the small ivory pointer *p*. Thus, when the mercury rises in the tube and sinks in the cistern, there would be a small space between the point of the ivory pointer and its reflection in the mercury; by means of the adjusting-screw S, the surface of the mercury can be raised until the two points seem to touch. The upper mirror *m* is used to avoid the error due to parallax.

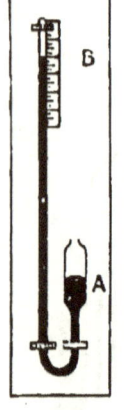

Fig. 72. Siphon barometer.

The **Siphon Barometer** is largely used. It is simple in construction, and consists of a U-tube, as shown in Fig. 72. The shorter limb, A, is open to the atmosphere, the longer is closed, and, as in the Torricellian tube, there is a vacuum above the mercury. When the mercury rises at B, it falls at A, and the "height" is the *difference of level* of the mercury in the two limbs. In other well-known forms, the U-tube is uniform in bore throughout, and carries at its open end, A, a small balanced piston or float, the rise or fall of which gives motion to a finger or pointer, and the indications are read off on a graduated dial.

Boyle's Law.

The volume of a given mass of gas varies inversely as the pressure, if the temperature be kept constant.

APPARATUS.—A Boyle's Law tube fixed to convenient stand; mercury.

EXPERIMENT 47.—Into a Boyle's Law tube (Fig. 73) a small quantity of mercury is poured, and the adjustment is carefully made so that the surface of the mercury in the two limbs is in the same horizontal line A A. The pressure of the air in both limbs is now the pressure of the atmosphere. Mercury is added at the open end C until the column of air in the shorter limb is reduced to one-half. The height of the column of mercury A F which is necessary to do this is found to be about 30 inches above the level of the mercury in the shorter limb.

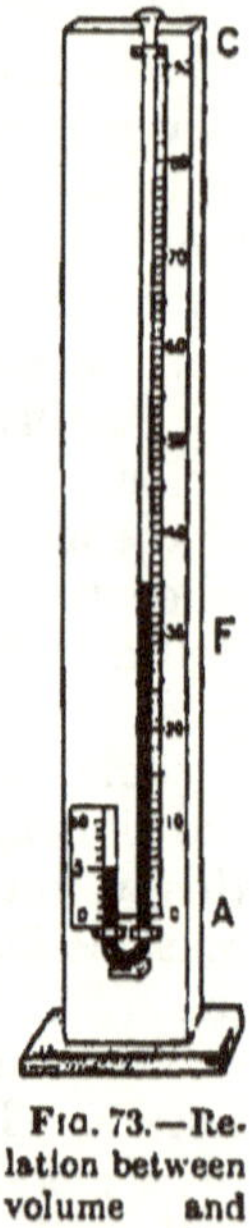

Fig. 73.—Relation between volume and pressure of a gas — Boyle's Law.

If, as shown in Fig. 73, scales are attached to both limbs, the volume of air in the shorter limb can be subjected to various pressures by altering the height of the column in A C. In this manner the law may be verified. Thus, if *v* is the observed volume of air when the pressure is p, and v_1 is the volume when the pressure is p_1, then the law states that $p \times v = p_1 v_1$, if the temperature remain constant; or, the product of the pressure into the volume is constant if the temperature be kept constant.

Of the many sources of error in the form shown in Fig. 73, the principal are: (a) the difficulty experienced in removing the air which enters with the mercury; (b) the rise in temperature if the mercury is not put in very carefully; (c) the difficulty in the initial adjustment. The apparatus is also inconvenient to use.

A better and more convenient form is shown in Fig. 74. A straight glass tube and a glass globe (or a glass tube of same bore may be used) are connected together by means of a strong india-rubber tube. The tube B is fastened by means of brass clips to a vertical stand C D. The globe A may be supported either by means of a retort-stand and ring, or, as shown in Fig. 74, a ring to carry the globe may be clamped on to the stand C D. In either case, the globe or bottle can easily be raised or lowered and secured in any position. Before using the *tube*, it *should be calibrated*. This can be done by detaching the rubber tube and weighing the tube when empty, also when *filled with mercury* to a height corresponding to the *first division* on the scale; the difference of weight should be noted. then the tube *filled to the second division* and again weighed. In this manner the weight of mercury between any two divisions can be ascertained. These, when plotted on squared paper and a fair curve drawn through, will enable the reading for any intermediate division to be made.

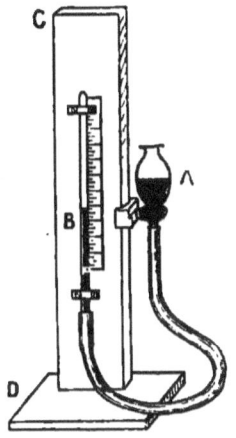

FIG. 74.—Apparatus to show relation between volume and pressure of a gas.

In the apparatus shown in Fig. 74, the pressure of the enclosed air, by lowering the globe sufficiently, may be made less than atmospheric. In this manner *Boyle's Law for lower pressures than the atmosphere* may be verified.

EXERCISES.

1. How does it appear that air is (a) heavy, (b) elastic, (c) fluid?
An air-tight globe of thin glass contains air: it is placed in the receiver of an air-pump; what will happen when the air is gradually withdrawn from the receiver, and why?

2. State Boyle's Law.

A certain quantity of air (A) has a volume of 70 cubic inches under a pressure of 25 inches of mercury; another quantity (B) has a volume of 80 cubic inches under a pressure of 35 inches of mercury: the temperatures being the same, find the ratio of A to B.

If the weight of A were 10 grains, what would be the weight of B?

Summary.

Air has weight; due to its weight it exerts pressure. This pressure at or near the earth is about 14·7 lbs. per sq. in., or roughly, 15 lbs. per sq. in.

Barometer, an instrument which indicates the pressure of the atmosphere.

Some of the barometers used are the **Torricellian, Siphon, Fortin's**.

Boyle's Law: If the temperature be kept constant, **the volume** of a given mass of gas is **inversely proportional to its pressure**; or, if p denote the pressure, v the volume, then the law is $p \times v = $ constant.

CHAPTER VI.

REPRESENTATION AND MEASUREMENT OF FORCES— PARALLEL FORCES AND CENTRE OF GRAVITY.

Force *is any cause which changes, or tends to change, the state of rest or motion of a body.* Force occurs in many different forms, and thus receives different names, such as *tension, pressure, friction, gravity,* etc.

Representation of a Force.—In order to determine a force completely, we must know (1) its *point of application,* (2) its *magnitude,* (3) *direction or line of action.* All these may be represented by a *straight line.* Thus one end of a line may represent the point of application, the number of units in its length denote the magnitude, and the direction, or better, the *sense* of the force may be indicated by an arrow-head on the line. In this convenient manner any quantity which is *directional*—such as a force, a velocity, an acceleration, etc.—may be represented by a straight line.

To denote the unit of weight any convenient scale may be used. Thus, if a *gram* be the unit of weight, and be represented by a line 1 cm. in length, then a length of 10 cm. would represent 10 grams; or if $\frac{1}{2}$ inch represent 1 pound, then a length of $2\frac{1}{2}$ inches would denote 5 pounds, etc.

Measurement of Force.—Every one is familiar with indiarubber cord, which, under the action of a pull, becomes longer, but returns to its former state when released. Bodies which possess this property of returning to their original state after being changed in form are called *elastic* bodies. A spring, when extended or compressed, will return to its original length when released, and is said to be *elastic.*

Force measured by Tension.—A force may be measured by means of the extension of a spiral spring. Thus a known weight,

78 REPRESENTATION AND MEASUREMENT OF FORCES—

W, applied to the spiral spring shown in Fig. 75, would stretch it through a certain distance. *As the elongation of an elastic body is proportional to the stretching force*, if the spring were perfectly elastic double the weight would stretch it to double the distance; and when the weight was removed, the spring would return to its original length, if the elongation produced by the weight had not been too great. Thus the elongation produced, which measures the force applied, can easily be ascertained by means of the two pointers, pp, and the graduated scale shown in Fig. 75.

There are several forms of **spring balance** in general use. In that shown in Fig. 76, a pointer, p, is made to move along a graduated scale; so that when a weight, W, is applied to the hook H, the indication on the scale gives the magnitude of W. Conversely, graduations on the scale can be obtained by applying known weights, and marking the corresponding positions of p for each weight.

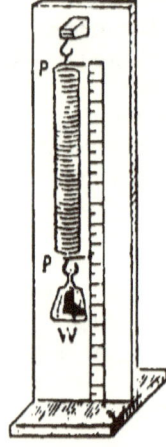

Fig. 75.—Measurement of weight by spiral spring.

Newton's Law states that the force of attraction between two bodies is directly proportional to their masses, and inversely proportional to the square of the distance between them. The direction of the force is in the line joining the centres of the bodies. At the poles, the distance from the surface to the centre of the earth is 13 miles less than at the equator, and by Newton's Law the nearer the centre the greater the pull. Thus a mass of iron weighing 190 ounces at the equator would weigh 191 ounces at the poles. For the same reason, the weight of a body is greater at the sea-level than on the top of a mountain. If any other attractive force is made to act on the body, the apparent weight may be increased or diminished. Thus the apparent weight is increased or diminished when a powerful magnet is brought below or above the scale pan containing the mass of iron.

Fig. 76. Spring balance.

The weight of a body, which is the earth-pull on the body—or, as it is called, the *force of gravity*—can be estimated by means of an ordinary balance (Fig. 77). The weight so obtained will be the same at any place on the earth's surface. If a spring balance,

PARALLEL FORCES AND CENTRE OF GRAVITY. 79

such as is shown in Fig. 76, be used, on account of the variation in the force of gravity it would be necessary, for accurate reading, that the balance be only used at the particular latitude where the graduations were made.

Another method of estimating **force** is by **momentum gained or lost per second.** *Newton's second law of motion* is, "*Change of motion is proportional to impressed force, and takes place in the direction in which the force acts.*" As rendered by Professor Maxwell, this becomes, "**The change of momentum of a body is equal to the impulse which produces it, and is in the same direction.**" **Momentum** or "quantity of motion" of a body is *the product of the mass and the velocity of a body.* The mass may be expressed either in pounds or in grams, and the velocity in feet or in centimetres per second. Example :—The momentum of a body whose mass is 10 lbs., moving at 30 feet per second, is 10 × 30 = 300 F.P.S. units. If the mass of the body had been 8 ounces, its momentum would have been $\frac{1}{2}$ × 30 = 15 F.P.S. units. If the mass of the body be 100 grams, and moving at 30 cm. per second, the momentum is 3,000 C.G.S. units. Or, if F denote the force, t the time in seconds during which the force F is acting, v and u the velocity of a mass, m, at the end and at the beginning of the time, t, then $Ft = m(v - u)$.

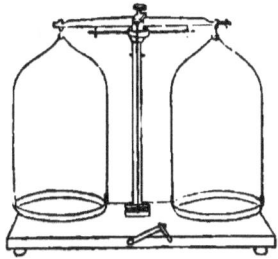

FIG. 77.—Balance.

The product Ft is called the impulse of the force, and $m(v - u)$ is the change of momentum. Hence we have **impulse = change of momentum.**

The change of momentum is produced by *force*, and the *measure of a force* is the *change of momentum* which it produces. If the mass remains constant, the rate of change of the velocity per second is called **acceleration**; hence if f denote acceleration, then **force = (mass) × (acceleration),** or $F = mf$.

Unit of force produces unit acceleration in unit mass.

Unit of momentum = (unit mass) × (unit velocity).

This would be in the F.P.S. system *a mass of 1 pound moving with a velocity of 1 foot per second;* in the C.G.S. system *1 gram moving at 1 centimetre per second.*

From the equation $F = mf$ we have at once a very useful means of estimating a force by the number of units of acceleration it can produce in a body of mass m. Hence *force is estimated by*

80   REPRESENTATION AND MEASUREMENT OF FORCES—

the gain or loss of momentum per second. As a special or particular case of $F = mf$, we have $W = mg$—that is, the weight of a body is g times the mass of the body. F is the force in poundals, m the mass, and W the weight of a body in pounds.

Velocity, which may be either uniform or variable, is the **rate of motion,** or the rate at which a body moves. It is said to be *uniform* when *equal spaces* are passed over *in equal intervals of time,* and *variable* when *unequal spaces* are passed over *in equal intervals.* Thus, if a body be moving at a uniform speed of 10 feet per second, in 4 seconds it will move through a space of 4×10 or 40 feet, or $S = vt$, where S is the space passed over by a body moving with uniform velocity, v, for a time, t.

When the velocity is **variable,** the velocity at any instant is the space through which it would pass in the unit time, assuming the velocity which it has at the instant considered was maintained during the unit of time. The **average velocity** $= \dfrac{\text{space described}}{\text{time taken in the journey}}$. Thus, if an express train starting from rest from a station reaches another station 200 miles distant from the former in 4 hours, the average velocity would be 50 miles per hour, or $\dfrac{50 \times 5280}{60 \times 60} = 73\tfrac{1}{3}$ feet per second.

Acceleration.—The express, starting from rest, gradually increases its speed. The *rate* per unit time at which the speed is altering is called the acceleration.

Acceleration is rate of change of velocity, and may be either uniform or variable. When the action is to increase the velocity of the body, the acceleration is said to be positive; and when it diminishes the velocity, it is said to be negative, or retardation.

If f denote the uniform acceleration of a body, then the velocity at any time t would be $v = ft$. Thus if a body moves from rest with a uniform acceleration of 5 feet per second, every second, the velocity at the end of 4 seconds is $5 \times 4 = 20$ feet per second. If the body had an initial velocity of 20 feet per second, the velocity at the end of 4 seconds would be $20 + 20$, or $20 - 20 = 40$, or 0. If the acceleration is positive, the velocity is 40 feet per second; if negative, the body is reduced to rest. This may be expressed by writing $v = u \pm ft$, where u denotes initial velocity of body, and v the velocity at the end of time, t.

The **space described** is equal to *the product of the average velocity and the time.* Or if s denote the space, v the velocity at the end and u the velocity at the beginning of the time,

PARALLEL FORCES AND CENTRE OF GRAVITY. 81

t, respectively, then the *average velocity* with *uniform* positive acceleration, f, is $\frac{1}{2}(v + u) = u + \frac{1}{2}ft$; ∴ $s = ut + \frac{1}{2}ft^2$.

For uniform retardation we obtain $s = ut - \frac{1}{2}ft^2$.

Hence
$$v = u \pm ft \quad \ldots \ldots (1)$$
$$s = ut \pm \tfrac{1}{2}ft^2 \quad \ldots (2)$$
and substituting from (1) in (2), $\quad v^2 = u^2 \pm 2fs \quad \ldots (3)$
(3) may be written as $\quad v^2 - u^2 = 2fs \quad \ldots \ldots (4)$

When the body starts from rest, the initial velocity is zero, or $u = 0$. Hence the equations become $v = ft$, $s = \tfrac{1}{2}ft^2$, and $v^2 = 2fs$. If the acceleration be that due to the force of gravity, then we write g for f. $\quad v = gt$, $s = \tfrac{1}{2}gt^2$, $v^2 = 2gs$.

Components and Resultants.—When two forces act at a point along the same straight line, their effect is the same as that of a force equal to their algebraic sum. The two or more forces which are acting at the point are called *components;* the single force equal to their sum is called the *resultant.* Thus, in Fig. 78, if the forces OA and OB, of 7 and 4 units respectively, act upon the point O, the resultant would be a force of three units acting from O in a direction OA; if the force OB acted in the same direction as OA, the resultant would be equal to 11 units acting from O towards A.

FIG. 78.

When two forces, not in the same line, acting on a body in the same plane, meet at a point, the resultant will act at the same point in the same plane, and be intermediate in direction to the two forces or components. The force equal in magnitude but opposite in direction to the resultant is called the *equilibriant;* the *components* and *equilibriant* form a system of forces in equilibrium, or balancing each other.

APPARATUS.—A board about 15″ × 11″, fixed in a vertical position, and carrying at each top corner a small wood or metal pulley about 3 inches in diameter, which should be well balanced, and made to rotate as easily as possible; three small scale pans (tin or brass), string, drawing-paper, drawing-pins, and weights.

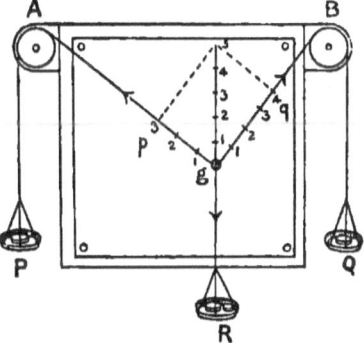

FIG. 79.—Apparatus to illustrate the parallelogram of forces.

A convenient method of attaching the pulleys is shown in Fig. 79, where AB is a strip of wood fastened to the back of the board, and projecting on each side, so that the pulleys will not rest against or touch the sides of the board. The three strings, which should be about 18 inches long, are conveniently attached to a small metal ring, g.

EXPERIMENT 48.—Fasten a sheet of drawing-paper to the board by means of drawing-pins; pass two of the strings over the two pulleys; put weights into all the pans. The total of the weights in any two pans must be greater than the weight in the remaining scale pan. When the scale pans are at rest, carefully mark the directions of the cords upon the paper, indicating the pull in each cord, which is the weight of scale pan together with the weight in it; also indicate by arrow-heads the direction or sense of each force. Alter the weights in the pans, and, if necessary, change the position of the paper; proceed as before, until three or four diagrams are obtained; remove the paper from the board, and draw lines through the points marked. Draw a triangle, having its three sides parallel to the directions of the forces in any of the above experiments, one side equal in magnitude to one of them on any convenient scale. Measure the other two sides of the triangle. These will be found to be nearly equal to the other two forces respectively. Repeat the construction for each diagram you have obtained.

If the weights P, Q, and R are in the ratio of 3, 4, and 5 respectively (say 30, 40, and 50 grams), the angle pgq (Fig. 76) is found to be a right angle, or 90°. When P, Q, and R are equal, the angle is 120°.

Owing to the friction of the pulleys and the weights of the strings, the results obtained by experiment are not quite accurate, but are of sufficient accuracy to show that the three forces keeping equilibrium are in amount equal to P, Q, and R.

The triangle which has its three sides equal in magnitude and parallel in direction to P, Q, and R respectively, is called the **triangle of forces.** *Hence, if three forces acting at a point be represented in magnitude and direction by the three sides of a triangle taken in order, they are in equilibrium.* The converse is also true: when three forces which act at a point are in equilibrium, the triangle drawn with its sides parallel to the directions of the forces respectively, will have those sides proportional to the forces to which each is parallel.

Although proved only for forces, the foregoing is true when applied to velocities and accelerations.

Parallel Forces.—When the forces do not meet at a point, but

PARALLEL FORCES AND CENTRE OF GRAVITY. 83

are parallel, they may act in the same or in opposite directions. In the former case they are called *like*, and in the latter *unlike* parallel forces.

In Fig. 80 a uniform bar or lath of wood, A B, about 36" long, $2'' \times \frac{7}{8}''$ in section, graduated into inches, is shown. A hole is made through the centre of gravity, and a steel spindle about $\frac{1}{8}$ inch in diameter, or a knife-edge, is inserted, the ends projecting about $\frac{1}{4}$ inch on each side. These may be supported by sheet-iron or other supports, which are fastened to a wooden stand. The whole is placed on a balance as shown, and the reading noted.

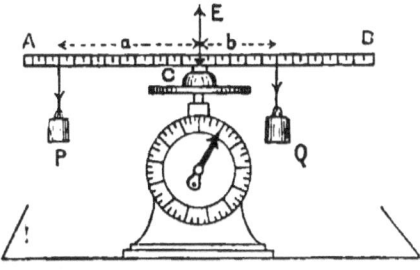

FIG. 80.—Apparatus to illustrate parallel forces.

EXPERIMENT 49.—Attach, as shown, a weight, P, to the bar by a loop of string at any convenient distance from the fulcrum (let this distance be denoted by a); also a weight, Q, larger than P. Find by trial the position of Q, so that the two forces balance. Let b be the distance of Q from the fulcrum. Prove that $P \times a = Q \times b$.

The products, $P \times a$ and $Q \times b$, are called the **moments of P and Q** respectively about the point C. The length, a, is sometimes called the *arm of P*, and b is called the *arm of Q*. The reading on the scale of the balance will be found to be $P + Q + w$ (where w = weight of bar and support).

Hence the equilibriant E, equal and opposite to the resultant, R, of two like parallel forces, P and Q, is $P + Q$, and the algebraic sum of the moments of P and Q about C is zero.

It should be carefully noted that the moment of P is the product of P into the perpendicular let fall from point E in the direction of P. Hence the **moment of a force about a point** is the **product of the force into the perpendicular from the given point on the line of action of the force.**

It is easy to show by changing the weights at P and Q that if the magnitude of P be doubled (the distances of P and Q from C remaining constant), the magnitude of Q must also be doubled, for equilibrium; or, for any increase in P, a corresponding increase must be made in Q. Thus **moment is proportional to force.**

In a similar manner, by changing the positions of P and Q along the bar, if the arm of P be doubled the arm of Q must

84   REPRESENTATION AND MEASUREMENT OF FORCES—

also be doubled (the magnitudes of P and Q remaining constant); and for any alteration in the position of P, a corresponding alteration in the position of Q must be made. Hence **moment is proportional to arm.**

Instead of two, three or more forces may be applied to either side of the bar, and the principle verified as in the former case.

Let E denote the force equal and opposite to R. If we consider the two unlike parallel forces E and P acting on the bar A B, the distance between them being denoted by a, the resultant is in magnitude $E - P = Q$, parallel to but acting in the opposite direction to E, and at a distance b from the line of action of E such that $Q \times b = P \times a$.

Couple.—In the case of two unlike parallel forces, the resultant is in magnitude equal to the difference of the two forces. Hence when the forces are equal and unlike parallel forces not acting in the same straight line, no resultant (or single force) can be found to balance. Instead, we obtain a *turning moment* called a **couple**, and this can only be balanced by an equal and opposite couple. The moment of the couple is **the product of either force into the perpendicular distance between them.**

An example of couples is obtained when equal forces are applied to the two handles of a book-press, etc.

APPARATUS.—The wood bar from last experiment; pulley and cord; weights.

EXPERIMENT 50.—Balance the beam as before, and attach two unlike parallel forces to the bar, which can readily be done by passing a string over a pulley as shown in Fig. 81, the pulley being attached to any convenient support. If the distances of the two equal forces from the fulcrum be equal, and each equal to a, the *moment of the couple* will be $P \times a + P \times a = P \times 2a$.

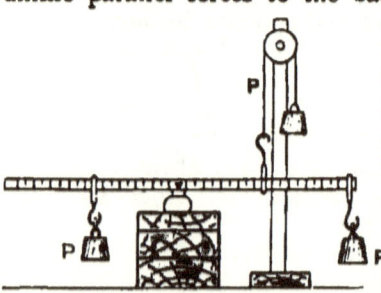

FIG. 81.—Experiment to illustrate a "couple."

Now attach a third and equal weight to the bar so as to produce equilibrium. Let b be the distance of this third force from the fulcrum; then $P \times 2a = P \times b$, $\therefore b = 2a$.

Note that the couple $P \times 2a$ in positive direction is balanced by an equal couple $P \times b$ in negative direction.

PARALLEL FORCES AND CENTRE OF GRAVITY. 85

If more convenient, a Salter's cylindrical balance may be used, as shown in Fig. 82; as before, R is found to be equal to P and $Q+w$, where w = weight of lever; also $P \times a = Q \times b$.

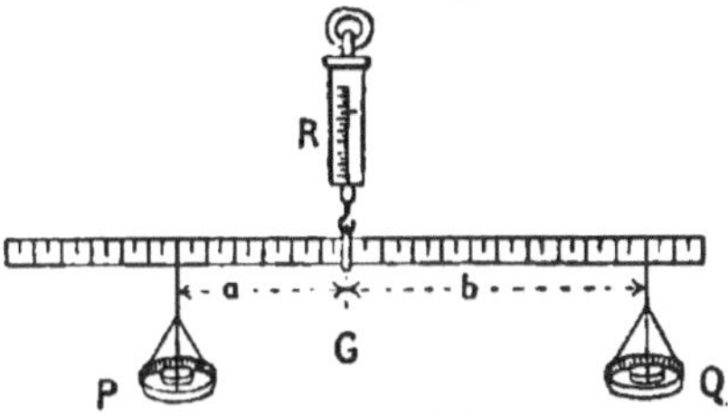

In each case the weight of scale pan is included. When the centre of gravity of the lever is at the fulcrum, the weight of the lever has no moment about that point. If G be the centre of gravity of the lever, and the point of support be on one side of G (Fig. 83), then if w denote the weight of the lever, acting at G, the distance of G from R being c, the weight, Q, acting at a distance, b, is balanced by P, acting at a distance, a; and when the lever is balanced, $P \times a = w \times c + Q \times b$. Hence $w = \dfrac{P \times a - Q \times b}{c}$. Thus indirectly we can find the *weight of the lever*, and the result obtained can be checked by actual weighing.

FIG. 82.—Parallel forces.

In the form shown, the bar A B is called a **lever**. A lever is defined as *a rigid bar which is capable of motion about a fixed point, or support, called the fulcrum*. A familiar example of a lever with equal arms is the balance, shown in Fig. 77.

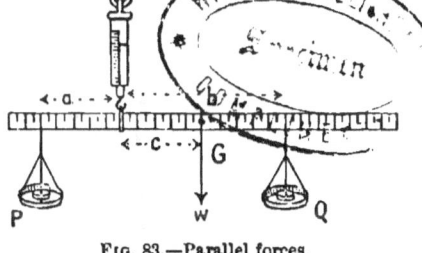

FIG. 83.—Parallel forces.

APPARATUS.—Wooden bar, as in last experiment; spring balance, scale pans, and weights.

EXPERIMENT 51.—Place any known weight, P, at a distance from the fulcrum denoted by a. Take a weight, Q, two or three times as great as P, and find a point on the bar so that the two weights, P and Q, balance; measure the distance of Q from the fulcrum, and call it b. Repeat the experiment, using different values of P and Q, and varying the distances of P and Q from the fulcrum. Tabulate thus:—

P.	Distance a.	$P \times a$.	Q.	Distance b.	$Q \times b$.
....	....	....	....	....	....
....	....	....	....	....	....
....	....	....	....	....	....

You will find that the product, $P \times a$ in column 3, will agree with the corresponding values of $Q \times b$ in column 6.

Centre of Gravity.—It is possible to balance some bodies on a knife-edge, such as the edge of a steel scale, or on a cylinder (for which a pencil may be used). When the body is balanced, there will be as much of its weight on one side the support as there is on the other.

We have found that when two like parallel forces are acting on a body, we can obtain a point about which the forces will balance, and the resultant which is equal to the sum of the two forces acts through the point. In a similar manner the resultant of any number of like parallel forces, which in magnitude is equal to their sum, can be found; its point of application is such that the **sum of the moments of all the forces on one side the point is equal to the sum of the moments on the other side.** Or it may be expressed as follows:—

The point of application is such that the *moment of the resultant about any point* is *equal to the algebraic sum of the moments of the components about the same point.*

The earth-pull or force of gravity on each particle of matter of which a body is composed constitutes an indefinite number of parallel forces. There is for every body a centre of weight, called the centre of gravity, or centre of mass, through which the resultant of all these parallel forces may be supposed to act; and if this point be supported, the body will rest in any position. This centre may usually be obtained either by experiment or by calculation.

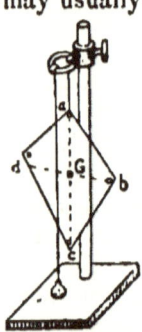

Fig. 84.—Centre of gravity.

APPARATUS.—Several pieces of cardboard, zinc, or thin wood, cut into various regular shapes, such as a triangle and a parallelogram, and also into irregular shapes; small lead ball and fine string, to form a plummet.

EXPERIMENT 52. — Determine the centres of gravity of the various shaped pieces of cardboard supplied to you. To do this, make small holes in any two angular points, as a and b (Fig. 84); suspend the figure from a, and in front of the figure hang a plummet from the same support, and draw on the face of the figure a vertical line $a\ c$. The centre of gravity of the face lies in this line.

Suspend the figure from any other point, such as b, so obtaining the line $b\ d$; the intersection of these lines gives G, the centre of gravity of the face $a\ b\ c\ d$. If the figure be uniform in thickness, the centre of gravity is at the point of intersection at a point half-way through the thickness. The construction can be verified by *balancing the figure on the point* G, on a pin placed in a block of wood.

The same result would be obtained by balancing the given figure on the edge of a steel scale. When balanced, the line along which the figure is resting is marked, the centre of gravity being at some point in the line. If when balanced in any other position another line be marked, the intersection of the two lines, as before, will give the centre of gravity required. In any regular polygon— that is, square, pentagon, hexagon, etc.—the centre of gravity is the point equidistant from all the angular points of the figure. In a uniform rod the centre of gravity is at its middle point. For a tapering rod, such as a pointer, the centre is nearer to the thicker end, and can be found by balancing on the edge of a scale. In a triangle the centre of gravity is two-thirds the line joining any angular point to the middle point of the opposite side, measured from the angular point.

APPARATUS.—Wood or cardboard triangle; square pentagon; semicircle, etc.

EXPERIMENT 53.—Find the centre of gravity in each case, and prove the rule.

APPARATUS.—Brass wire 2 or 3 mm. diameter, or skeleton wire, cube or tetrahedron, finer wire, thread.

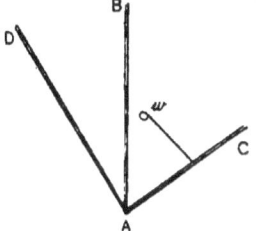

FIG. 85.—Centre of gravity of a wire frame.

EXPERIMENT 54.—To obtain the centre of gravity of a wire framework. In Fig. 85, three wires, about 2 or 3 mm. diameter and 12 or 14 cm. in length, are soldered together at A. To one of them, as A C, a piece of fine wire is attached. The framework is hung up by one of the corners, B, to any convenient support, and the wire, w, is bent so as to embrace loosely the string of the plummet; a piece of thread or fine string may be attached to indicate the direction. It is next suspended from another point, C, and the point where the vertical line through C cuts the thread or string is determined; this is the centre of gravity required.

In a similar manner the verticals through the centre of gravity of the skeleton cube or tetrahedron can be obtained—the former at the point of intersection of the diagonals, the latter at a point one quarter the line joining the middle point of the base to the opposite vertex.

88 REPRESENTATION AND MEASUREMENT OF FORCES—

Summary.

A force may be represented in magnitude, direction, and position by a straight line.

If **two or more forces** act upon a particle in the same or in contrary directions, and in the same straight line, their effect will be the same as that of a single force equal to their sum or to their difference respectively.

Momentum is the quantity of motion of a body, and is measured by the product of its mass and velocity.

Unit of momentum, the product of unit mass and unit velocity.

Acceleration, the force causing motion ÷ mass moved.

Like Parallel Forces.—The resultant is equal to the sum of the forces, its point of application being at some definite point in the straight line joining the forces.

Unlike Parallel Forces.—The resultant is equal in magnitude to the difference of the forces; its point of application lies in the straight line joining the given forces produced and nearer to the greater. In each case the position is such that the algebraic sum of the moments of the forces about the point is zero.

Centre of gravity is that point about which a body or any system of bodies may be balanced.

Exercises.

1. The velocity of a body is increased uniformly in each minute of its motion by 66,000 yards a minute; by how many feet a second is its velocity increased in each second?

If the acceleration of a body's velocity, due to the action of a certain force, is 55 in feet and seconds, what is it in yards and minutes?

2. (a) A particle moves along a straight line, and its velocity is uniformly accelerated: at a certain instant it was moving at the rate of 800 yards a minute, and after 6 seconds it is moving at the rate of 55 feet a second. What is the acceleration in feet and seconds?

(b) If the mass of the particle is 12 oz., how many poundals are there in the force to which the acceleration is due?

3. Define an absolute unit of force when a foot, a second, and a pound are taken as units. By what name is this unit of force commonly known? How many such units are there in the force which, acting for 3 seconds on a particle whose mass is 5 oz., communicates to it a velocity of 360 yards a minute?

4. What distinction is usually made between the "mass" and the "weight" of a body?

What is the gravitation unit of force, and how do you derive it from the British absolute unit?

If a mass of 2 lbs., moving uniformly at the rate of 10 feet a second, be acted on by the British absolute unit of force for 2 seconds, find the final velocity.

5. A body whose mass is 3 lbs. acquires in 5 seconds a velocity of 80 feet a second; how many poundals are there in the force which, acting in the direction of the motion, caused the body to acquire this velocity?

6. A body whose mass is 12 lbs. is found to gain a velocity of 15 feet a second when acted on by a constant force (P) for 3 seconds; find the number of poundals (or British absolute units of force) in P. What ratio does P bear to the force exerted by gravity on the body?

7. A body whose velocity undergoes constant acceleration has at a certain instant a velocity of 22 feet a second; in the following minute it describes 10,320 feet: find the constant acceleration.

8. The velocity of a body is increased uniformly in each second by 20 feet a second; by how many yards a minute will its velocity be increased in one minute?

9. Define a poundal (or British absolute unit of force).

If a force of five poundals acts on a mass of 10 lbs. in the direction of the motion, what velocity would it impart to the mass in 3 seconds?

PARALLEL FORCES AND CENTRE OF GRAVITY.

10. Draw an equilateral triangle A B C, and produce B C to D, making C D equal to B C; suppose that B D is a rod (without weight) kept at rest by forces acting along the lines A B, A C, A D: given that the force acting at B is one of 10 units acting from A to B, find by construction (or otherwise) the other two forces, and specify them completely.

Mention the points that go to the specification of a force.

11. State the rule for the composition of two velocities.

Draw a square A B C D and a diagonal B D. A particle moves along A B with a given velocity; find the velocity that must be impressed on it at B to make it move from B towards D with a velocity equal to that with which it was moving along A B.

12. If a man be walking with velocity 6 through rain descending vertically with velocity 20, determine graphically in magnitude and direction the velocity of the rain relative to the man.

13. Draw a triangle A B C, whose sides, B C, C A, A B, are 7, 9, 11 units long. If A B C is the triangle for three forces in equilibrium at a point P, and if the force corresponding to the side B C is a force of 21 lbs., show in a diagram how the forces act, and find the magnitude of the other two forces.

14. (a) A B C is a triangle whose sides, B C, C A, A B, are 7, 4, 5 units long; at A two forces act—one of 8 units from A to C, and one of 10 units from A to B. Draw a straight line through A to represent their resultant in all respects, and state the number of units of force in the resultant.

(b) A ball moving at the rate of 10 feet a second is struck in such a way that its velocity is increased to 12 feet a second, and the direction of the new velocity makes an angle of 45° with that of the old velocity; find by construction the velocity imparted by the blow, and its direction.

15. Draw a line A B, take a point C in it, and make an angle B C D of 40°. A particle moves from A to C with a velocity 20; at C its velocity is changed to 30 along C D; find by construction drawn to scale the magnitude and direction of the velocity impressed on the particle at C.

16. State the condition of the equilibrium of three forces acting at a point called the triangle of forces.

Show by a diagram drawn to scale the lines along which three forces of 13, 12, and 5 units must act if they are in equilibrium, and find from the diagram the angle between each pair of forces.

17. Enunciate the proposition known as the triangle of forces.

One end of a string is attached to a fixed point A, and after passing over a smooth peg B in the same horizontal plane, sustains a weight of P lbs. A weight of 50 lbs. is now knotted to the string at C, midway between A and B. Find P so that in the position of equilibrium A C may make an angle of 60° with A B.

18. Two forces act at a point along two given lines respectively; state how their resultant can be found by construction.

Draw an angle A O B of 120°, and draw O C within the angle A O B so that A O C may be 45°. If a force of 100 units acts from O to C, find (by construction or otherwise) its components along O A and O B.

19. Draw an equilateral triangle A B C. A body moves from B to A with a velocity of 20 feet a second; find the velocity that must be impressed on it at A to make it move from A to C with a velocity of 20 feet a second.

20. Draw a square A B C D. A particle moves along A B with a velocity 10; it is made to move along B C with a velocity 20; find the magnitude and direction of the velocity that must be impressed on it at B.

21. Draw an equilateral triangle A B C. A force of 10 units acts from A to B, and one of 15 units from A to C; find their resultant by construction. Also find what their resultant would be if the force of 15 units acted from C to A.

22. A body slides down a smooth inclined plane, the height of which is 10 feet and the length 100 feet; find (a) the acceleration of the body's velocity while sliding; (b) the velocity which the body acquires in sliding from the top

REPRESENTATION AND MEASUREMENT OF FORCES—

to the bottom of the plane; (c) the time it takes, starting without initial velocity, to get from the top to the bottom ($g = 32$).

23. Given the formula $s = \frac{1}{2} f t^2$, state the meaning of each symbol in it. If the formula is to be applied to the case of a falling body, and f is given equal to $32\cdot19$ or (approximately) 32, what supposition is implied as to units of time and distance?

Neglecting the resistance of the air, find the time taken by a body in falling from the top of the Eiffel Tower, 300 metres high. ($N.B.$—A metre = $3\cdot281$ feet.)

24. Taking for granted the formula—
$$s = Vt + \tfrac{1}{2} f t^2,$$
state the meaning of each symbol in it. If a body falling freely describes a distance of 30 yards in half a second, what was its velocity at the beginning of the half-second? ($g = 32$.)

25. A point moves in a straight line, and its velocity undergoes a constant acceleration; state how the acceleration is measured.

If the point acquires in 15 seconds a velocity of 300 miles an hour, what is the acceleration in feet and seconds? What ratio does this acceleration bear to the acceleration of the velocity of a body falling freely?

26. A body is thrown upward from the top of a tower with a velocity of 48 feet a second; find where it will be at the end of 4 seconds ($g = 32$).

Write down the formula or formulæ by means of which you answer this question, and state what it means (or they mean).

27. A body falling freely has a velocity V at a certain instant; in 3 seconds from that instant it falls through a distance a, and in 6 seconds from that instant it falls through a distance b. If the ratio of a to b equals that of 4 to 13, find V.

28. State the rule for finding the resultant of two like parallel forces.

What is meant when two parallel forces are said to be "like"?

A uniform rod A B, 6 feet long, rests on two points in a horizontal line, 5 feet apart; one of the points is under A. If the weight of the rod is 20 lbs., find the pressure on each point.

29. A straight line A B represents a rod 10 feet long, supported horizontally on two points, one under each end; C is a point in A B, 3 feet from A; what pressure is produced on the points A and B by a weight of 30 lbs. hung at C?

What additional pressure is exerted on the points of support if the rod is of uniform density and weighs 20 lbs.?

30. Let A B represent a horizontal line 10 feet long, and F a point in it 6 inches from A; suppose that A B is a lever that turns on a fulcrum under F, and carries a weight of 50 lbs. at B; if it is kept horizontal by a fixed point above the rod, 5 inches from F and 1 inch from A, find the pressure on the fulcrum and on the fixed point.

31. A rod or lever is capable of turning freely round a fixed point or fulcrum, and is acted on by a force at each end; putting the weight of the lever out of the question, state the relation which must exist between the forces when the rod stays at rest.

Draw an equilateral triangle A B C, and let B C represent a weightless lever acted on at B by a force of 7 units from A to B, and at C by a force of 9 units from A to C. If the lever is at rest, find, by construction or otherwise, the position of the fulcrum; find also the magnitude and direction of the pressure on the fulcrum.

32. A weightless rod A B rests horizontally on two points under A and B 14 feet apart; it carries a weight suspended from a point X, which causes a pressure of 3 units on A, and of $7\frac{1}{2}$ units on B; find the distance of X from A.

33. Equal forces act on two bodies whose masses are M and m; at the end of a second the former is moving at the rate of 10 miles an hour, and the latter at the rate of 110 feet a second; find the ratio of M to m. State the physical principle that justifies your answer.

PARALLEL FORCES AND CENTRE OF GRAVITY. 91

34. A body moves along a straight line; its mass is 5, and its velocity is 7; at the end of three units of time its velocity is 19. Find (a) the change that has taken place in its momentum; (b) what will be the change in the momentum at the end of each unit of time, if the force that acts on the body is constant.

Mention a familiar instance of a body moving in a straight line under the action of a constant force.

35. Define momentum.

If a gun weighing 70 tons fires its 650 lbs. projectile with a velocity of 1,700 feet per second, find its velocity of recoil.

Compare the kinetic energy of recoil with the kinetic energy of the projectile as it leaves the muzzle of the gun. Why would you not expect to find them equal?

36. A lamina of uniform density is in shape a parallelogram; where is its centre of gravity situated?

If the lamina weighs 4 lbs., and a particle weighing 1 lb. is placed at an angular point, where is the centre of gravity of the whole situated?

37. Particles whose masses are 2, 5, 2, 3, are placed in order at the angular points of a square; show in a diagram the position of their centre of gravity, and find its distance from the particle whose mass is 5.

38. In what sense may an area be said to possess a centre of gravity?

A B C D is a plane quadrilateral having the sides A B, C D parallel. Determine, graphically or otherwise, the position of its centre of gravity.

39. Draw an equilateral triangle A B C, and let particles whose masses are 5, 6, 7 be placed at A, B, C respectively; find by construction the centre of gravity (G) of the particles, and note what part A G is of A C.

Mention one property of the centre of gravity of a body.

40. State where the centre of gravity is situated in (a) a cylinder, (b) a parallelogram, each being of uniform density.

41. A B is a rod of uniform cross section, 12 feet long; it is made up of two rods of equal length, A C and C B, firmly joined at C; they are of different materials, so that A C weighs 3 lbs., and C B weighs 2 lbs. (a) Find the distance from A to the centre of gravity of the rod; (b) find also what weight must be placed at B that the centre of gravity of the whole may be at C.

42. A B C D is a rectangle, and A C is one of its diagonals; four equal particles are placed one at each angular point; where is their centre of gravity? If, other things remaining the same, the particle at C is moved to A, where is now the centre of gravity of the four particles?

43. State how to find the centre of gravity of two particles whose masses are given, and which are placed a given distance apart.

If the masses are 5 and 7 units, and they are placed 3 feet apart, how far is their centre of gravity from each of them?

CHAPTER VII.

WORK AND ENERGY—PRINCIPLE OF WORK—SIMPLE MACHINES.

Work.—If a force acting on a body causes it to move into a different position, or changes the form or size of the body by overcoming the resistances which oppose the change, the force is said to do *work* on the body; thus *work is done by a force when its point of application is moving in the direction in which the force acts.* The amount of **work** is measured by the product of the force F multiplied by the distance S, through which it acts, or the resistance multiplied by the space through which it is overcome; or, work = FS. In the case of a weight w, raised to a vertical height h, the resistance is due to the action of gravity, and the *work done* is the *product of the weight multiplied by the height h*, or wh: this is called the **potential energy** of the weight.

The **unit of work is the product of unit force multiplied by unit distance.**

In the F.P.S. system, the unit distance is a foot; but the unit force may be the poundal, pound, ton, etc., and the work done is said to be in *foot-poundals, foot-pounds, foot-tons,* etc. In the C.G.S. system, a *dyne* is the force which, acting on a mass of one gram, produces a velocity of one centimetre per second. The **unit of work** is the work done by a dyne through the space of one centimetre, and is called an **erg**.

Energy is the capacity or ability to do work which a body possesses in virtue of its position or its motion.

Measure of Potential and Kinetic Energy.—The energy of position of a weight w raised to a height h is wh, and is called the potential energy of the body; if allowed to fall, the velocity due to a height h is given by $v^2 = 2gh$; hence $wh = \dfrac{wv^2}{2g}$.

WORK AND ENERGY—PRINCIPLE OF WORK. 93

The expression on the right is called the **kinetic energy** of the body; it is *numerically equal to the work done on the body, and also represents in foot-pounds the amount of work which must be done on the body to bring it to rest.*

If m denote the mass of a body, then since $m = \dfrac{w}{g}$, the **kinetic energy** *of a mass, m*, moving with velocity, v, is $\tfrac{1}{2}mv^2$.

This very important result may also be obtained in a more general manner as follows:—

A *constant* force f, acting on a body or mass m, changes the velocity from u to v in a time t, during which the body moves through a distance s; then if f denote the acceleration produced,

$$v^2 - u^2 = 2fs,$$
or $\tfrac{1}{2}mv^2 - \tfrac{1}{2}mu^2 = mfs = FS.$

If the body start from rest, u is zero, and hence $FS. = \tfrac{1}{2}mv^2$.

Other familiar instances of *potential energy* are, a compressed or wound-up spring, etc.; and of *kinetic energy*, in the motion of a projectile or a fly-wheel, etc.

In addition to the two important forms of energy just referred to, there are other forms, such as **heat energy, strain energy,** and also **chemical energy, electrical energy, sound,** and **light.** Many illustrations of all the forms of energy just referred to are easily obtainable; only a few can be referred to here.

In a *steam-engine* the heat given to the water converts it into steam, which may be at a pressure considerably greater than that of the atmosphere; this, when allowed to press against the piston in a cylinder, may, by means of suitable mechanism, be made to do the work of driving machinery, etc. In gas and oil engines an explosive mixture is compressed behind a piston, and when fired great heat is developed, and in consequence great pressure is obtained, which can be made to do useful work in a variety of forms. In a similar manner work is done when a spring, being wound up, is bent or compressed: the spring acquires potential energy, and in running down may be made to do useful work, such as driving the mechanism of a clock or watch, etc. When a current is sent through a wire, as shown later on in Experiment 183, the wire is heated, and may be made red-hot or incandescent, giving light.

Conservation of Energy.—Careful experiments of great accuracy have been repeatedly made, and it has been shown conclusively that the disappearance of energy of one kind gives rise to an equal amount of another; thus in the case of potential energy and kinetic energy we have found that the one is numerically equal to the other, or the sum total remains constant for all positions of the body.

Power is the rate of doing work, and it is evident that some suitable **unit** must be selected. An agent doing **33,000 foot-pounds of work** in **one minute**, or 550 foot-pounds per second, is said to work at the rate of **one-horse power**. This is the unit adopted in the F.P.S. system.

There are various contrivances called **machines** by which the force acting on a body can be modified in amount, in direction, or in both together.

The **principle of work** states that *the work expended on a machine is equal to the work done by the machine, together with the work lost in frictional resistances.* Thus, if the work done by a machine in any given time is 1,000 foot-pounds, and it is known that 250 foot-pounds are lost in frictional resistances, then the work in foot-pounds expended on the machine must be 1,000 + 250, or 1,250. It is often convenient to employ a small force moving through a large distance, which can be made to exert a comparatively large force by moving through a small distance. Thus, if E and W denote two forces acting on a machine, the first denoting the applied force or effort, and the latter the weight raised or resistance overcome; L and l the distances moved through by E and W respectively, in the same time; then the *work expended* on the machine is E × L, and *the work done by the machine* is W × l; hence if F denote the work lost in friction, then

$$E \times L = W \times l + F.$$

If we neglect friction, then

$$E \times L = W \times l$$
$$\text{or } \frac{E}{W} = \frac{l}{L}.$$

This is sometimes expressed by, *What is gained in force is lost in speed.*

The ratio of W to E is called the **mechanical advantage**. It should be carefully noted that no force is created by the machine, but in every case there is a loss due to frictional and other resistances. The work actually done by a machine is called the **useful work**, the amount of **extra work** which has to be done on a machine to enable it to do this work is called the **lost work**, and

SIMPLE MACHINES. 95

the sum is called the **total work**; the *efficiency of the machine is the ratio of the useful to the total work.*

$$\therefore \text{efficiency} = \frac{\text{useful work}}{\text{total work}}.$$

The simple machines are—(1) the **lever**, (2) the **pulley**, (3) **inclined plane** and **screw**.

Inclined Plane.—If a *smooth* cylinder C be prevented from rolling down a *smooth,* inclined plane (Fig. 86) by a force E acting in a direction parallel to AD, the three forces which keep C at rest are—the *force* E, the *reaction* R, and the *weight* W.

If to any convenient scale a line BH is made equal to W, and HF and BF intersecting in F are drawn parallel to E and R respectively, the lengths of HF and BF, when measured on the same scale, give the amounts of the force E and the reaction R.

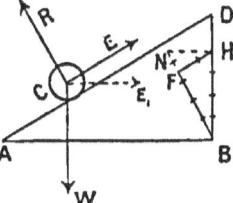

FIG. 86.—Inclined plane.

It will be seen that the triangle BFH is similar to the triangle ABD; if the triangle BFH were laid on the triangle ABD so that F falls on B and the line FH coincides with a portion of BD, then BF would coincide with BA, and BH would be parallel to AD.

Hence
$$\frac{E}{W} = \frac{FH}{BH} = \frac{BD}{AD} = \frac{\text{height of plane}}{\text{length of plane}}.$$
$$\therefore E \times AD = W \times BD.$$

In a similar manner, when the direction of E is *horizontal* or parallel to the base AB, drawing HN parallel to the base AB and producing BF to meet it in N, HN denotes the force E_1 to the same scale that BH denotes W; also BN to the same scale is the pressure on the plane.

$$\therefore \frac{E_1}{W} = \frac{HN}{HB} = \frac{BD}{AB} = \frac{\text{height of plane}}{\text{base of plane}},$$
$$\text{or } E_1 \times AB = W \times BD.$$

APPARATUS.—A piece of board AB, about 24″ × 6″ × 1″, is fastened to the lower edge of a fixed vertical board ABDH (Fig. 87). A piece of board AC, 20″ × 6″ × 1″, is attached to AB by a hinge, *h;* by means of a tightening screw or a pin, the piece AC, moving on the graduated scale shown, can be set at a desired inclination.

A roller carried by a light brass frame moves freely along the inclined surface AC, being held in position by a string attached at one end to the brass frame, and passing over a pulley F (which can be fixed in any position on the vertical piece BD), and has at its extremity either a hook to which weights may be attached, or a small scale pan as shown.

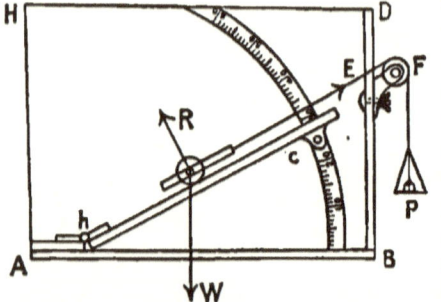
Fig. 87.—Apparatus to illustrate inclined plane.

Instead of the roller, if more convenient, a small sledge may be used; this should run as frictionless and as easily as possible on four small pulleys.

EXPERIMENT 55.—(a) Fix the plane at any convenient inclination; add weights at scale pan P until the roller, when started (which is easily effected by rapping the board), moves up the plane at a uniform speed.

Draw a triangle, having one side vertical and equal to W, the other two sides parallel to the directions of R and E respectively. Measure the value thus obtained for E, and compare with the value obtained by experiment.

Alter the inclination of the plane, and repeat the experiment.

(b) When the scale pan falls through a distance l, measure the corresponding height h, to which W is raised; then the work done along the plane $E \times l$, and the work done in raising W, or $W \times h$, are *nearly equal in each case;* they are not quite equal on account of the work wasted on friction, etc.

Tabulate your results as follows:—

W	E	h	l	Loss due to friction ($El - Wh$).

(c) Repeat the experiment when the pull E is horizontal. The pull can be applied horizontally by using a simple wire frame wide enough to clear the sides of the plane, and adjusting the position of the pulley F. In this case, when the roller moves up the plane, the direction of the string will be altered, but the inclination will remain approximately the same for a small motion.

(*d*) Ascertain the value of E, just sufficient to cause motion; draw a triangle, having one side vertical and equal to W, the other two sides at right angles to the plane and parallel to the base respectively. Measure the value of E so obtained, and compare with that obtained by experiment, and from the formula

$$E \times b = W \times h.$$

(*e*) Alter the inclination of the plane and position of pulley F, and repeat the experiment until five or six results are obtained.

Tabulate thus :—

W	E	h	l	Loss due to friction (Eb − Wh).

In the **simple machines** the **mechanical advantage** is most easily obtained by the **principle of work**. As any complex machine is usually made up of a number of simple machines, the relation between P and W is found by multiplying together the mechanical advantage obtained from each simple machine of which it is composed.

The lever has been already considered, but the relation between E and W can easily be obtained by the principle of work.

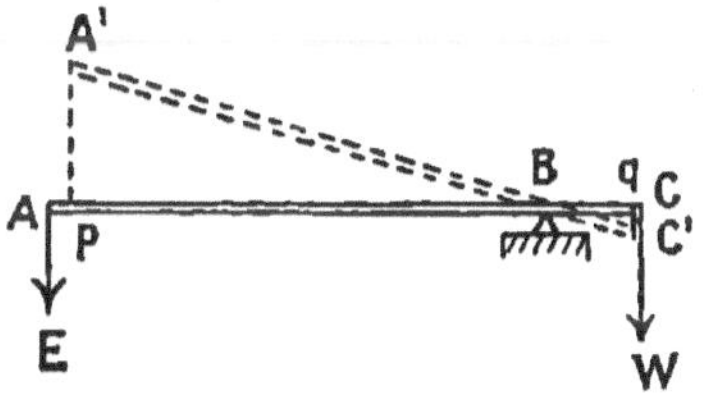

Thus in Fig. 88, when the lever AC moves into a new position A'C' (the angle through which the lever moves is supposed to be small). Draw the vertical lines A'p and C'q, intersecting the line AC in p and q respectively.

FIG. 88.—Lever.

The work done on $E = E \times A'p$.

The work done on $W = W \times C'q$, and these being equal, $E \times A'p = W \times C'q$;

$$\therefore \quad \frac{E}{W} = \frac{C'q}{A'p}.$$

98   WORK AND ENERGY—PRINCIPLE OF WORK—

But, as will easily be seen from the similar triangles $A'BA$ and CBC',

$$\text{the ratio } \frac{C'q}{A'p} = \frac{BC'}{BA'} = \frac{BC}{BA};$$

hence $E \times BA = W \times BC$; or $E \times a = W \times b$, where a and b denote the lengths of the arms AB and BC respectively.

In the lever shown in Fig. 88, the moment of E about the fulcrum, or $E \times AB$, is equal to the moment of W about the same point, or $W \times BC$; but when the lever is inclined to the horizontal, the moment of E is $E \times Bp$, and the moment of W is $W \times Bq$; hence in accurate measurements by the lever, as in **testing machines**, etc., arrangements are made by which the lever can be kept horizontal during the operation.

Pulley.—In the case of a pulley, which may be considered as a lever with equal arms, for any downward motion of one section E a corresponding amount of motion of W occurs in an upward direction. Thus the pulley is useful in changing the direction of a force, the leverage remaining constant. The pulley usually consists either of a disc, as shown (Fig. 89), or of a rim which may be grooved so that a string or rope can rest in the groove, and which is connected to a cylinder of metal at the centre by "arms." It is made to rotate as easily as possible on a spindle passing through its centre, and supported by a suitable block b.

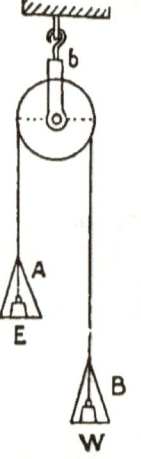

Fig. 89. Simple pulley.

Let E denote the load on one side and W that on the other. The two forces for equilibrium are equal in magnitude. To cause E to move downwards, a slight addition to it must be made to overcome the friction of the spindle and the rigidity of the rope.

If E' is the amount required at B to make the two loads move at a uniform rate when started from rest, then $E' - E$ is a measure of the friction. Instead of E' we may use W to denote the load which will give uniform motion.

APPARATUS.—Small fixed pulley and scale pans, weights, and sheet of squared paper.

EXPERIMENT 56.—Fix the pulley to any convenient support; put equal weights in the scale pans at A and B: it will be found that if an impulse is given to one of the scale pans, it will only continue to move for a short time, and through a small distance,

SIMPLE MACHINES. 99

before friction brings the system to rest. Weights are added to one scale pan until, when a slight impulse is given, the system moves at a fairly uniform rate; the added weights are required to overcome the friction of the apparatus. Call the weight in one scale pan E and in the other W (in each case including the weight of the scale pan): W − E is the amount required to overcome the frictional resistances. Tabulate as follows:—

E	W	Effect of friction or (W − E).

Alter the weight E and find the corresponding amount in W; repeat the experiment, using different weights until about ten results are obtained.

Plot the values of E and W − E on squared paper, thus obtaining a series of points; the curve which lies most evenly through the points will be found to be a straight line. Deduce the law connecting E and F by taking any two points in the straight line and substituting the values in the equation $F = aE + C$, where a and c are constants.

The following example will serve to illustrate the above, and similar methods may be followed in all the simple machines:—

E	W	F (effect of friction).
9·5	10·1	·6
11·5	12·2	·7
13·5	14·3	·8
15·5	16·4	·9
17·5	18·5	1·0
21·5	22·6	1·1
23·5	24·7	1·2
27·5	28·9	1·4

Plot the values tabulated in columns 1 and 3 as co-ordinates of points; thus, when E is 9·5, F is ·6. This is shown by point a (Fig. 90). When E is 11·5, F is ·7; when E is 27·5, F is 1·4,

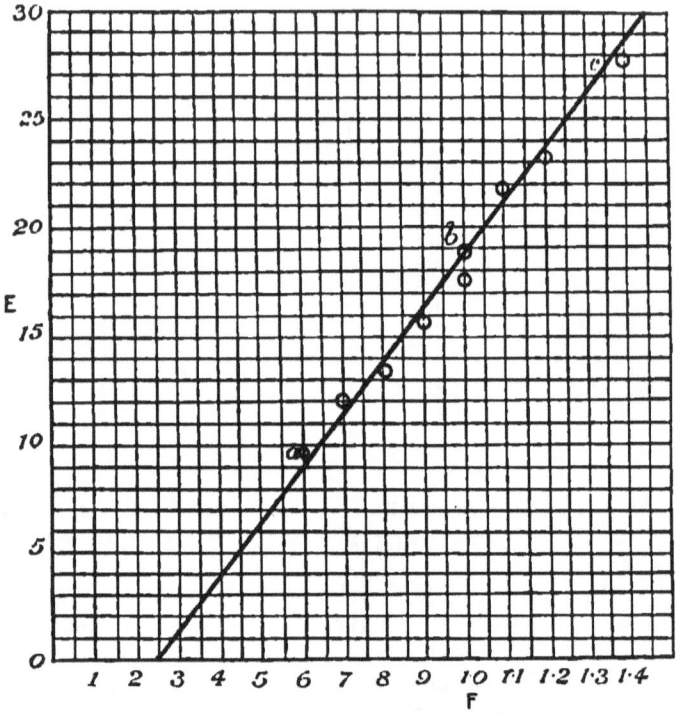

FIG. 90.—Exercise in plotting.

etc. When all the points are plotted, by means of a piece of black thread obtain the line ac, which lies most evenly amongst the points. To obtain the *law of the line*, the equation is $F = aE + c$, where a and c are constants. To determine the constants, we may proceed as follows:—

At the point a, $F = ·6$ and $E = 9$.

At another point b, $F = 1$ and $E = 18·5$; hence using these values, we have

$$
\begin{aligned}
1·0 &= a \times 18·5 + c \\
·6 &= a \times 9·0 + c \\
\hline
·4 &= \phantom{a \times {}} 9·5\ a, \text{ by subtraction.}
\end{aligned}
$$

$\therefore\ a = ·042.$

SIMPLE MACHINES. 101

Substituting this value in the first equation, we obtain
$$·6 = 9 \times ·042 + c,$$
from which $c = ·2668$.
Hence the *law required* is $F = ·042 \, E + ·2668$.

It is convenient at the outset to consider the *simple machines* as frictionless; the above will show the method adopted when friction is taken into account.

Single Movable Pulley.

APPARATUS.—Fixed pulley from last experiment, also a movable pulley to use with it; scale pans and weights.

In this case, a cord or rope is fastened to any convenient point C, and passes round a movable pulley B, and over a fixed pulley A; by means of a hook or a scale pan at the end of the rope, as shown in Fig. 91, a load E can be applied. If there were no friction, E (the pull in the string) would be $\frac{1}{2}W$; but on account of friction, E is greater than $\frac{1}{2}W$.

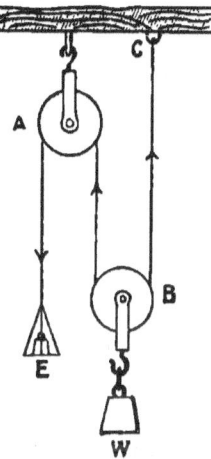

Fig. 91.
Single movable pulley.

EXPERIMENT 57.—Ascertain by experiment the load required at E to give (when started from rest) a slow uniform motion. Tabulate as in last example.

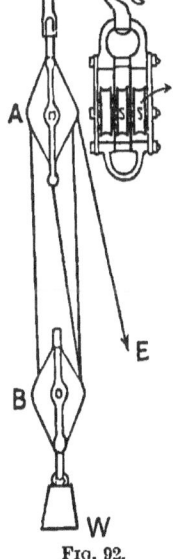

Fig. 92.
Pulley blocks.

Pulley Blocks, or *Blocks and Tackle*.—Blocks and tackle so called consist of two blocks, A and B (Fig. 92), each block containing a number of sheaves or pulleys: in Fig. 92 pulley blocks containing three sheaves are shown. A rope passes round each pulley in turn: one end of the rope is fastened to the upper block, as shown; to the other the pull E is applied. The weight W hangs as shown from the lower block. To ascertain the relation between E and W, a scale pan or a hook may be attached to the end of the rope.

When friction and the weight of the lower block are neglected, the pull in the rope is the same throughout and equal to E, and

WORK AND ENERGY—PRINCIPLE OF WORK—

the force acting on the lower block is equal to W, and is as many times E as there are portions of rope proceeding from the lower block. In Fig. 92 there are six portions of rope leaving the lower block; ∴ E = ⅙W.

Or by *principle of work*. If the weight be raised one inch, then each of the six cords will be shortened one inch, or E must fall six inches; hence the velocity ratio is 6, and E × 6 = W × 1, ∴ E = ⅙W. More exactly, the weight of the lower block should be added to W before proceeding to find E.

If there were no friction, the *velocity ratio* just obtained would also be the *mechanical advantage;* but there is friction, and instead of E being ⅙W, it would have to be *greater* to give a slow, steady motion to the weight. The difference between W and 6E would be the effect of friction for the load W, where W includes the weight of the lower block.

Tabulate as follows:—

E	W	F (effect of friction).	Efficiency.

Plot as co-ordinates of points on squared paper the values of E and W, and deduce the law connecting them; thus, $W = aE + c$.

Also plot the values of E and F, or W and F, and deduce the law $F = aE + c$.

The following example will illustrate the method:—

Blocks and tackle—3 sheaves in each block. Plot on squared paper and obtain the *laws*:—

$$W = 5E + 1.30; \quad F = 1.1E + 66.84.$$

E	W	F (friction).	Efficiency.
124	498	246	·67
134	549	255	·67
144	598	266	·69
154	640	284	·69
164	690	294	·70
174	736	308	·70

SIMPLE MACHINES. 103

The Screw.—As already mentioned (p.30), the screw may be looked upon as an inclined plane wound round a cylinder.

The **mechanical advantage of the screw** can be obtained most easily by the *principle of work*.

In the simple screw-jack shown in Fig. 93, the weight to be raised is placed on the head H. The square-threaded screw which works in a nut N can be made to advance or recede by means of a lever passing through either of the two holes shown. A turned projecting piece on the end of the screw fitting into a corresponding recess in the head H allows the screw to rotate, the head H rising and falling with the screw, but not rotating.

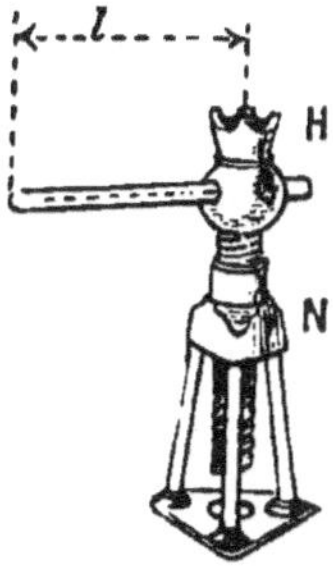

If l is the length of the lever (or the distance from the axis of the screw to the point where the force E is applied), then in one rotation of the lever the force E moves a distance $2l \times 3\cdot1416$; the head H and W moves a distance equal to the *pitch;* hence we have $E \times 2l \times 3\cdot1416 = W \times$ pitch,

or $E \times 2\pi l = W \times p$ $\left\{\begin{array}{l}\text{where } p = \text{pitch} \\ \text{and } \pi = 3\cdot1416 \text{ or } 3\tfrac{1}{7}\end{array}\right\}$.

FIG. 93.—Screw-Jack.

The same result is obtained in the case of the screw-press shown in Fig. 94. The force W denotes the pressure exerted on the book or object in the press.

APPARATUS.—The upper part of the screw-jack shown in Fig. 93 is replaced by a grooved pulley (which may be made of

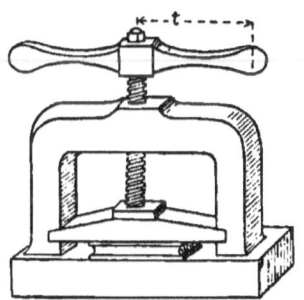

FIG. 94.—Screw-press.

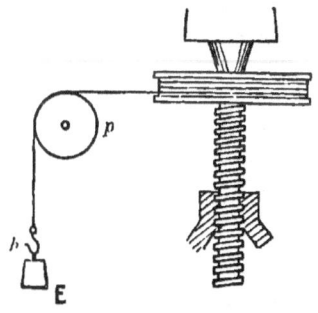

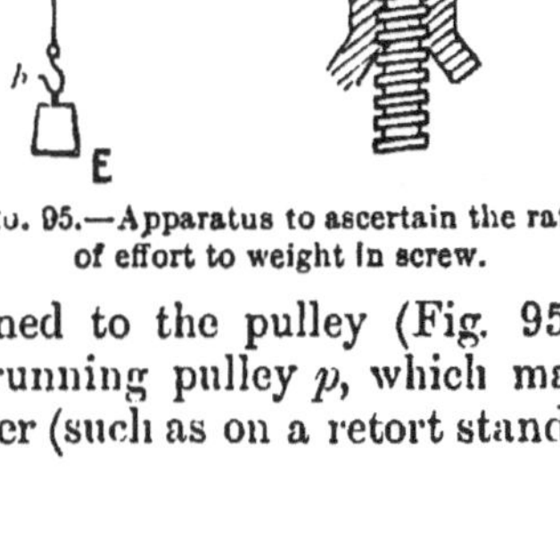

FIG. 95.—Apparatus to ascertain the ratio of effort to weight in screw.

wood). One end of a cord is fastened to the pulley (Fig. 95), the other passes over a small, easy-running pulley p, which may be supported in any convenient manner (such as on a retort stand).

104 WORK AND ENERGY—PRINCIPLE OF WORK—

EXPERIMENT 58.—First obtain the velocity ratio. This will be the ratio of the circumference of the pulley to the pitch of the screw. (The circumference is easily ascertained by measuring the length of cord required to pass once round the pulley.)

If there were no friction, then one pound at the hook h would maintain steady motion while r lbs. were raised at A (where r denotes the velocity ratio). On account of frictional resistances, the force or effort required is greater; hence find the amount required to keep up a steady motion; call this E; then the difference between rE and the load at A, which we may denote by W, is the friction for that load.

Repeat the experiment, using different loads at A.

Tabulate thus:—

E	W	r E-W (effect of friction).

Friction.

When two surfaces are in contact, the force parallel to their common surfaces which exists between them, and which resists the movement of one surface over the other, is called the *force of friction*. This force acts *in a direction opposite to that in which motion takes place.*

The friction between two surfaces depends on the pressure producing it, and also on the state of the surfaces in contact. It is measured experimentally by the force necessary to maintain uniform motion of one surface over the other.

APPARATUS.—A board A B, about 3 or 4 feet in length, 6 inches wide, and 1 inch thick, fastened to a table (or supported on brackets against a wall) in a horizontal position. The board should be planed so as to present a fairly uniform surface. At one end of the board fixed to the table a small pulley is placed; this should turn as freely as possible. The other end of the board is fastened by means of a *hinge h* to a fixed piece A.

Several smaller boards of pine, oak, etc., called "sledges," have fixed to them a small hook to which a cord is attached. The cord passes over the pulley, and terminates in a hook or

SIMPLE MACHINES. 105

scale pan. The pulley should be placed so that the cord is parallel to the surface of the board.

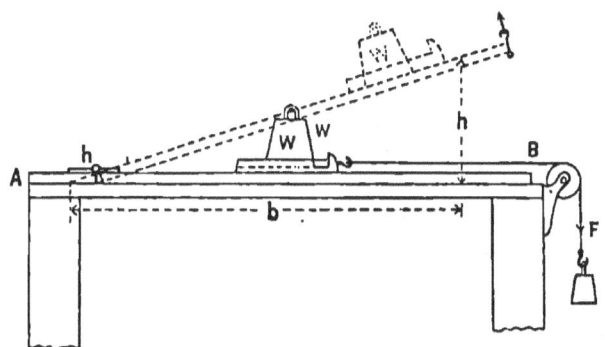

FIG. 96.—Friction.

EXPERIMENT 59.—Choose any sledge, and let the weight on it together with its own weight = W; then add weights at P until the sledge moves uniformly. To do this, the board must be rapped or the sledge given a slight impulse. Let F be the pull in the cord when the sledge moves uniformly, beginning with the weight of the sledge; increase the load upon the sledge by weights—that is, 4 lbs., 7 lbs., 14 lbs., etc. Find in each case the value of F, and tabulate :—

W	F	$\mu = \dfrac{F}{W}$

As shown in the third column, the ratio of $\dfrac{F}{W}$ is usually denoted by μ, where μ is the *coefficient of friction*.

That friction is proportional to pressure is shown by reference to the first two columns, giving values of W and F.

Plot the values of W and F as co-ordinates of points on squared paper, and it will be found that the curve which lies most evenly amongst the points is a straight line. Deduce the law $F = aW + c$.

The friction between two surfaces in contact is less when there is motion than when they are at rest. Thus in the last experiment, if the force at F is gradually increased until the sledge starts from rest, the value of F is greater for any value of W than that tabulated; and it will be found that when motion occurs, the velocity is not uniform, but increases as the body moves along the surface.

Angle of Friction.

APPARATUS.—The board and sliders used in previous experiments.

EXPERIMENT 60.—Taking the same slider, and using the same values of W as before, raise one end of the board so that the surface makes an angle with the horizontal, and find the angle θ when a motion given to the slider is kept up uniformly; fix the board in this position; drop a plumb-line from the raised end of the board (this vertical line will, with the horizontal surface of the table-top and the inclined surface of the plank, make a right-angled triangle); measure the height h and base b of this triangle (Fig. 96). Then $\frac{\text{height}}{\text{base}} = \mu$, from which μ can be found by calculation; or a jointed two-feet rule can be inserted so that two edges of the rule fit the angle. The angle may be marked on a sheet of paper, and assuming any convenient length as base, the height can be measured, and the value of μ can be calculated.

Repeat the experiment, using different weights, and tabulate thus:—

W	$\frac{\text{Height}}{\text{Base}}$

Compare the mean value of μ with the mean value of μ obtained in the last experiment.

EXPERIMENT 61.—*Friction depends on the nature of the surfaces in contact.* This can be verified by using sliders of different materials; wood, cast or wrought iron, or glass may be used.

Results obtained can be compared thus. When the load W is the same, the corresponding values of F will differ from each other.

SIMPLE MACHINES.

	W					F			
Pine with grain.	Pine against grain.	Cast-iron.	Mahogany.	Glass.	Pine with grain.	Pine against grain.	Cast-iron.	Mahogany.	Glass.

EXPERIMENT 62.—*Friction is independent of area of surfaces in contact.* This can be shown by using two sliders of same material, the area of one being two or three times the area of the other, and finding μ in each case. It should be noted that the result obtained is only true within certain limits, and is not to be taken as correct in all cases.

SUMMARY.

Work is done by a force when the resistance of a body to change of position, or to change of shape or size, is overcome.

Unit of Work is the *product of unit force × unit distance.*

The unit force may be a poundal, pound, ton, etc., and the unit distance a foot, an inch, etc., and the work done is expressed as so many foot-poundals, foot or inch pounds, etc.

Energy is the capacity for doing work, and may be *Potential* or *Kinetic;* instances are furnished in *Heat,* in *Electricity,* in *Chemistry,* and in *Light* and *Sound.*

Kinetic Energy of a moving body is *one-half the product of the mass by the square of its velocity.*

Power is *rate of doing work.*

Principle of Work.—Work expended or given to a machine is equal to the useful work done by it, together with the work spent in frictional resistances.

Simple Machines—the *lever, pulley, inclined plane,* and *screw.*

EXERCISES.

1. Define a foot-pound and a horse-power.

A steam crane raises a weight of 5 tons uniformly through a height of 110 feet in 40 seconds; find at what H.-P. it is working.

2. A body whose mass is 10 lbs. moves in a straight line, and its velocity

is changed from 210 feet a second to 90 feet a second; find the numerical value of the change of its kinetic energy. State what units are used in your answer.

If the change is produced by a force equal to 3 lbs. weight ($g = 32$), through what distance does the particle move while its kinetic energy is undergoing the change?

3. A machine is contrived by means of which a weight of 3 tons by falling 3 feet is able to lift a weight of 168 lbs. to a height of 100 feet; find the work done by the falling body, and what part of the work is used up in overcoming the friction of the machine.

4. B is a point 50 feet above A; a body whose mass is 10 lbs. is thrown vertically upward from A; reckoning in foot-poundals, find its gain of potential energy on reaching B.

If at A its kinetic energy were 54,720 foot-poundals, find its kinetic energy at B, and hence its velocity at B.

5. A body whose mass is 15 lbs. is moving at the rate of 40 feet a second; how many foot-poundals of work can it do against a resistance, in virtue of its mass and velocity?

Define a foot-poundal of work.

6. Write down a formula for the work which can be done by a particle in virtue of its mass and velocity. In this formula what units must be used to obtain the work in foot-poundals?

A body whose mass is 10 lbs. moves from rest along a straight line under the action of a force of 80 poundals; find its velocity when it has described 50 feet.

7. Write down the formula for the kinetic energy of a particle of given mass moving with a given velocity.

If the mass of a particle is 1·8 ounces, and it is moving at the rate of 1,500 feet a minute, how many foot-poundals of work can it do against a resistance in virtue of its mass and velocity?

8. Define a unit of work and a foot-pound.

A steam-engine raises 250 gallons of water per minute from a depth of 880 feet; at what horse-power does it work? (*N.B.*—A gallon of water weighs 10 lbs.)

9. A body whose mass is 18 lbs. has a kinetic energy of 49 foot-poundals; find its velocity.

If the body were moving against a resistance of a third of a poundal, how far would it move before coming to rest?

10. Define a unit of work and a unit of power; give an example of each.

A man does 1,027,200 foot-pounds of work in 8 hours; what is his power, the units being foot-pounds and minutes?

11. Define horse-power, and express 1 horse-power in foot-poundals per second.

What horse-power is required to draw a weight of 1 ton up a smooth plane, inclined at 30°, at the rate of 20 feet per minute?

Express 1 horse-power in ergs per second.

12. The base of a cylinder has a diameter of 3 feet, and its height is 4 feet; the cylinder is of uniform density, and weighs 25 cwt. Find how many foot-pounds of work must be done in throwing it over.

13. A body weighing 100 lbs. is observed to be moving at the rate of 20 feet a second; assuming that it began to move from a state of rest, and that its motion was impeded by no resistance, how many units of work must have been done on it by the force that gave it the velocity? Give the answer (*a*) in foot-poundals; (*b*) in foot-pounds, assuming $g = 32$.

14. A number of men can each do, on the average, 495,000 foot-pounds of work per day of 8 hours; how many of such men are required to do the work at the rate of 10 horse-power?

15. A rod 7·5 feet long can move freely in a plane round one end; it is

acted on at right angles to its length by a force of 35 lbs.; find the number of foot-pounds of work done by the force in one turn. How many turns must it make a minute if the force works with 1 horse-power? ($\pi = 3\frac{1}{7}$.)

16. What is the horse-power of the engine which draws a train with a uniform rate of 45 miles an hour, against a resistance of 900 lbs.?

17. A body whose mass is 5 lbs. falls freely through 10 feet; what kinetic energy (in foot-poundals) does it acquire?

If, instead of falling freely, it were fastened by a fine thread to a mass of 3 lbs., and the thread were placed on a smooth point so that the mass of 5 lbs. has to draw the mass of 3 lbs. up, find the kinetic energy acquired by the mass of 5 lbs. in falling 20 feet ($g = 32$).

18. A particle moving from rest is acted on through 250 feet by a force of 9 poundals; find its kinetic energy, and its mass being 5 lbs., find its velocity.

19. AB is a rod, 20 feet long, that can turn freely round the end A; at B a force of 35 lbs. is applied at right angles to AB; the rod is allowed to turn six times; how many foot-pounds of work are done by the force?

20. A body whose mass is 20 lbs. moves in a straight line against a constant resistance R; at a certain point it is moving at the rate of 18 feet a second; after moving over 50 feet its velocity is reduced 10 feet a second; what part of its kinetic energy has it lost? What is the numerical value of R in poundals?

21. If the mass of a body is 15 lbs., and its velocity 12 feet a second, how many foot-poundals of work can it do against a resistance, in virtue of its mass and its velocity?

22. A body whose mass is 6 lbs. is moving at the rate of 8 feet a second; how many foot-poundals of work can it do against a resistance, in virtue of its mass and its velocity?

If it did 117 foot-poundals of work against a resistance, what would then be its velocity?

23. If a man can work at the rate of 210,000 foot-pounds an hour, how long would it take him to raise a weight of 10 tons through 150 feet, supposing him to be provided with a suitable machine?

24. Given the equation of work and energy for a particle—namely, $\frac{1}{2}mv^2 - \frac{1}{2}mV^2 = Ps$—state the meaning of each letter in the equation.

A body whose mass is 12 lbs. has its velocity changed from 5 feet a second to 11 feet a second; what number of foot-poundals of work has been done by the force to which the change is due?

25. A body whose mass is 10 lbs. is carried up to the top of a house 30 feet high; by how many foot-poundals has the change in position increased its potential energy? If it is allowed to fall, what number of foot-poundals of kinetic energy will it have when it reaches the ground? ($g = 32$.)

If the mass of a body is 24 lbs., and its kinetic energy is 75 foot-poundals, what is its velocity?

26. A body whose mass is 10 lbs. is capable of doing 605 foot-poundals of work; what is the velocity in feet per second?

CHAPTER VIII.

PENDULUM AND ATTWOOD'S MACHINE.

Pendulum.

A simple pendulum consists of a heavy particle suspended from a point by a weightless thread. Such an ideal form cannot be realized by any physical apparatus, but is obtained with sufficient accuracy for all practical purposes by using a fine thread fastened at one end to a suitable support, and carrying at its free end a small lead ball. The length of the pendulum l is the distance from the point of suspension to the centre of gravity of the ball.

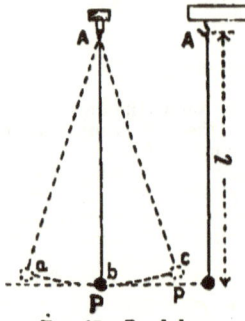

Fig. 97.—Pendulum.

Let A P (Fig. 97) represent such a pendulum. When pulled on one side of its mean position and let go, it will continue to swing to and fro through some such arc as abc. This arc, owing to the resistance of the air, etc., will gradually get less and less, until finally the pendulum comes to rest again at P.

Let t denote the time of a vibration from a to c; at a place where the acceleration g due to gravity = 32·2 ft. per sec. per sec.

$$\text{then } t = \pi \sqrt{\frac{l}{g}}.$$

The **period** is the time taken for a **complete vibration** from a to c and back again to a, and is twice the above, or $2\pi \sqrt{\dfrac{l}{g}}$.

The **amplitude** of the vibration is the extreme distance, ba or bc, on each side of the mean position. When the *amplitude* is small (not exceeding an angle of about 20°), the vibrations take place in nearly equal times, and are said to be **isochronous**; in other words, for a **small vibration the time is independent of the amplitude**.

The pendulum furnishes a good illustration of the exchange of *potential* for *kinetic energy*. When drawn to one side of its mean position, the potential energy is equal to the product of the weight of the ball P and its vertical height pc (Fig. 97). When released, this potential energy is gradually exchanged for kinetic energy, until at the lowest point, P, *nearly* all the potential energy is converted into kinetic energy; in other words, the velocity which is zero at the highest point increases gradually, and is greatest as it passes the lowest point, P. The velocity becomes less and less until it comes to rest at a height nearly the same as before, the slight loss referred to being due to frictional resistances; hence, when used in a clock, to maintain the arc of swing the mechanism is so arranged that a slight impulse is given to the pendulum at each swing or vibration of the pendulum.

A **Compound Pendulum** consists of a body of any shape supported on a horizontal axis and free to swing about that axis; the foot of a perpendicular let fall from the centre of gravity of the body to the axis of suspension is the **centre of suspension**. If the perpendicular be produced backwards to a point whose distance from the centre of suspension is equal to the length of a simple pendulum which oscillates in the same time as the compound pendulum, the point so obtained is called the **centre of oscillation**, the time of swing (which may be obtained experimentally by obtaining the mean of a large number of vibrations) being the same when suspended from either centre.

APPARATUS.—Lead ball; string.

EXPERIMENT 63.—Make a pendulum of any convenient length; pull it to one side until the amplitude is about 15° or 20°, and let it swing; observe the number of oscillations made per minute. Make the amplitude very small, and repeat the experiment. It will be found that the number of vibrations is the same as before.

To prove by experiment that **the time of oscillation is independent of the mass** and also of the **material** of the **pendulum**.

EXPERIMENT 64.—Hang two pendulums (large and small lead balls respectively) side by side; adjust the lengths until the two pendulums oscillate in the same time; when this occurs, the lengths will be found to be equal. Hence *the time is independent of the mass*. A similar result is obtained when one of the balls is lead and the other wood. Thus *the time is independent of the material.*

To find the law connecting the **length of a pendulum** *with its* time of oscillation, suspend the pendulum as before; measure as accurately as possible the distance from the point of support to the top of the ball, and add to this the radius of the ball to obtain the length of the pendulum.

APPARATUS.—Lead ball and fine thread.

EXPERIMENT 65.—Make a pendulum about 200 cm. long; observe and record the time of a large number of oscillations, say 50.

Repeat the experiment, but with a pendulum a quarter the length of the preceding one. It will be found that it vibrates twice as quickly. Observe and record a considerable number of vibrations.

Again alter the length of the pendulum, and proceed as before. Tabulate thus :—

Length.	Time of 50 Vibrations.	t, or Time of 1 Vibration.	t^2.	$\dfrac{l}{t^2}$

You will find that the numbers in the last column are either equal or nearly so.

Now $\dfrac{l}{t^2}$ or $\dfrac{\sqrt{l}}{t}$ could not remain the same unless the numerator and the denominator changed together, and in the same proportion.

Hence the **time of oscillation of a pendulum is proportional to the square root of the length of the pendulum**.

If the results obtained are plotted on a sheet of squared paper, taking the lengths of the pendulums as horizontal distances or abscissa, and value of t (or time of one oscillation) as vertical distances or ordinates, a series of points is obtained through which, if a curve (which lies evenly among the points) be drawn, the length of a pendulum which will vibrate in any given time—say 1 second—can be found. From the preceding experiment, knowing l and g, and having observed t, the time (in seconds) of a swing, the formula $t = \pi \sqrt{\dfrac{l}{g}}$ can be verified.

PENDULUM AND ATTWOOD'S MACHINE.

An approximate value of g at the place of observation may be obtained by measuring as accurately as possible the length, l, and the time, t, for any pendulum. It is advisable, to ensure accuracy, that the times of a large number of oscillations be taken.

EXAMPLE.—A lead ball 1½ inches diameter, length of string from point of suspension to centre of ball 83 inches, when set swinging was found to make 200 vibrations in 5 minutes. Find the value of g at the place of observation.

Since 5 minutes = 300 seconds, $\therefore t = \dfrac{200}{300} = \dfrac{103}{150}$,

and we know $t^2 = \dfrac{\pi^2 l}{g}$.

$$\therefore g = \dfrac{\pi^2 \times \dfrac{83}{12}}{t^2} = \dfrac{\pi^2 \times \dfrac{83}{12}}{\left(\dfrac{103}{150}\right)^2} = 32\cdot 188.$$

(The accurate value of $g = 32\cdot 198$.)

If in the formula $t = \pi \sqrt{\dfrac{l}{g}}$ we make $t = 1$, then l, the length of pendulum which will make a single vibration per second, is found to be 39·2 inches —the same result as obtained from the curve plotted in Experiment 61. Hence the length of a pendulum to vibrate in half-seconds would be a quarter the above, or 9·8 inches. If we take $g = 980$ cm./sec.^{2}, then $l = \dfrac{t^2 g}{4\pi^2} = $ **99·3 cm.**

Hence at a place where $g = 980$ the *period* would be 2 seconds, and each vibration would be 1 second, the length of the pendulum being 99·3 cm.

Attwood's Machine.

In order to measure the acceleration of a falling body, it is necessary to reduce the speed at which it moves. This is effected in *Attwood's machine* by increasing the mass moved without increasing the force of gravity. Any easy-running pulley with unequal weights, such as in Fig. 98, may be used. Many forms of the machine are to be obtained at a small cost, or can be easily constructed, and of sufficient accuracy to verify useful formulæ.

The Attwood's machine in its simplest

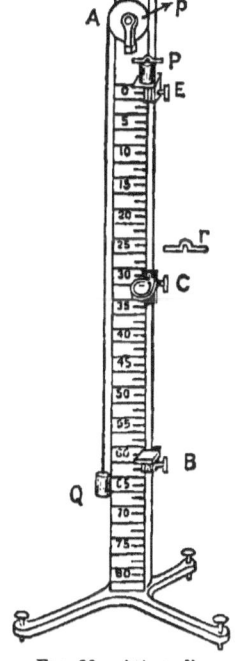

FIG. 98.—Attwood's machine.

form consists of a small grooved pulley, *p*, which is made to rotate as freely as possible. Two masses, P and Q, are attached to the ends of a fine string which passes over a fixed pulley, A. In addition to P and Q, small masses or weights, one of which is shown at *r* (Fig. 94), and which can be placed on the mass P, are provided: these are called *riders*. It will be seen that as the masses P and Q are equal and balance each other, the *moving force* is the weight of the rider *r*, and equal to rg if r is the mass. In this manner the acceleration due to gravity may be as small as required.

The fixed pulley A is carried by a pillar, as shown in Fig. 94, which is usually of wood, about 7 or 8 feet in height, having a scale graduated into inches or centimetres, so that the distance moved by the masses can be accurately estimated.

A small platform, B, and a ring through which P can pass easily, are provided; these can be clamped at any convenient position along the pillar.

In the more elaborate machine the pulley at A is made to turn on friction-wheels—that is, the axle rests on the circumferences of four light wheels, two on each side, and a seconds pendulum may also be attached to the back of the pillar.

The small rider *r* cannot pass through the ring C. When the platform at E, carrying P and *r*, is removed, motion begins. The rider *r* is left on the ring C as P passes through it. The weight P now moves over the distance C B with constant velocity. The time taken to move from E to C and from C to B is carefully noted; the acceleration from E to C is found from the formula

Moving force = mass moved × acceleration,

$$\therefore f \text{ or acceleration} = \frac{\text{moving force}}{\text{mass moved}}$$

$$= \frac{r}{2P + r} \times g.$$

From this the acceleration, *f*, can be found, and the value of *g* can be estimated.

As *f* and *t* are both known, the distance E C is obtained from the formula $S = \frac{1}{2} f t^2$. The distance can be measured and a verification of the formula obtained.

The ring C is clamped to the pillar in such a position that the under side of the cylindrical weight P is coincident with a division on the scale when the rider *r* is in contact with the ring.

As the rider *r* is left on the ring at C, the two masses are

PENDULUM AND ATTWOOD'S MACHINE. 115

moving at a uniform speed. Hence if t, the time taken from B to C, be determined (since $v = ft$, where v is the uniform speed of P), then $S = vt$. The distance BC or S being measured, the formula may be verified as before.

The friction of the pulley p can be allowed for by attaching a small weight to P of such an amount that when a slight impulse is given to P the system continues to move at a uniform speed.

To show that the acceleration of a given mass is proportional to the force acting on it.

EXAMPLE.—In an experiment the weights of P and Q were equal, and each 3,087 grains. When 56 grains were added to P, and a slight impulse given to P, the weights were found to move at a uniform rate, thus eliminating the effect of the friction of the pulley. With a rider placed on P, weighing 156 grains, the total mass moved would be $2 \times 3087 \times 56 + 156 = 6386$—the moving force 156 grains.

Let f denote the acceleration; then $f = \frac{156}{6386} g$.

Hence, using the formula $S = \frac{1}{2} f t^2$, the space described in 2 seconds from rest $= \frac{1}{2} \times 9\cdot4 \times 4 = 18\cdot8$.

Thus if the ring C be at a distance of 18·8 inches from E, the rider will be left on it at the end of 2 seconds; also P will go on from the ring C with a uniform velocity (of 18·8) acquired in 2 seconds, and the distance described in the next 2 seconds will be $2 \times 18\cdot8 = 37\cdot6$ inches; or if the distance from the ring C to the platform B is 37·6 inches, then P will reach it at the end of 4 seconds. The distance from E to B = 37·6 + 18·8 = **56·4**.

In a similar manner it was found that if a rider weighing 240 grains were used, the distances of C and B from E would be 28·6 inches and 85·9 inches respectively.

In the above experiments, to ensure accuracy, the mass of the pulley was estimated and allowed for; but the calculation cannot be given here, as it involves the consideration of moment of inertia.

SUMMARY.

The time of a vibration $t = \pi \sqrt{\dfrac{l}{g}}$.

Period is the time taken for a complete vibration, or $2\pi \sqrt{\dfrac{l}{g}}$.

Amplitude, extreme distance on each side of mean position. When amplitude is small, vibrations are *isochronous*.

Time of swing is independent of *mass* and material, but is proportional to length.

Attwood's machine, f or acceleration $= \dfrac{\text{moving force}}{\text{mass moved}}$.

EXERCISES.

1. For a given centre of suspension of a compound pendulum what is meant by the phrase "centre of oscillation"?
How can it be shown experimentally that these points are convertible?
Calculate the length of the seconds pendulum.
($N.B.$—$\pi = 3\cdot14159$; $g = 32\cdot1912$.)

2. Define a simple pendulum, a compound pendulum, and the centre of oscillation of a compound pendulum.

If a body suspended on a horizontal axis makes 35 small oscillations a minute, what is the distance from its centre of suspension to its centre of oscillation? ($g = 32$, $\pi = 3\frac{1}{7}$.)

3. In the case of a simple pendulum, the time of a small oscillation is given by the formula $2\pi\sqrt{\dfrac{l}{g}}$; what is meant by an oscillation? what by a small oscillation? What do the letters π, l, and g in the formula stand for?

A pendulum 8 feet long makes 38 beats a minute; what is the acceleration due to gravity at the place of observation?

4. If a compound pendulum makes 25 oscillations in a minute, what is the distance from the centre of suspension to the centre of oscillation? ($g = 32$; $\pi = 3\cdot1416$.)

5. Find how many beats a pendulum 40 inches long would make in 24 hours at a place where the acceleration due to gravity is 32 in feet and seconds. (*N.B.* $\sqrt{60} = 7\cdot745967$.)

6. What is a seconds pendulum? If we could have a simple pendulum in a place where $g = 30$ in feet and seconds, what would be the length of a seconds pendulum? ($\pi^2 = 9\cdot87$.)

7. Two particles, whose masses are P and Q, are connected by a fine thread. Q is placed on a horizontal table, and P hangs over the edge; there is no friction. If P is allowed to fall, find the acceleration.

If P is 1 lb., and Q is 7 lbs., find how long it would take P to fall through the first 3 feet of its descent.

8. Two particles, whose masses are P and Q, are connected by a fine thread which passes over a smooth pulley (as in Attwood's machine), so that when P descends it draws Q up. The bodies are allowed to move from a state of rest, and it is found that P falls through 9 feet in the first 3 half-seconds of its motion. Find the ratio of the mass of P to that of Q.

CHAPTER IX.

SOUND.

Sound Waves.

APPARATUS.—A wooden trough about 4 or 5 feet long by 6 inches deep; a small block of wood; india-rubber tube, $\frac{1}{4}$-inch bore and about 12 feet long, filled with sand; some beeswax and iron-filings.

EXPERIMENT 66.—Fasten one end of the tube to the ceiling. Holding the tube rather taut, give the end a slight shake; the motion travels up the tube to the ceiling. No permanent change has taken place in the state of the tube; the shake given to the tube caused a motion to run along it which is called a *wave motion*, or shortly a wave.

EXPERIMENT 67.—Half fill the long rectangular trough with water. Place small pieces of cork on the surface of the water. Alternately depress and elevate a block of wood placed at one end; by this means the surface of the water is thrown into waves. It will be seen that although the waves travel *forward* and *backward*, the pieces of cork simply move up and down, and these indicate the real motion of the water—that is, the particles of water are simply moving up and down although a wave is travelling along the surface. In a similar manner, in the last experiment the particles of sand and tube simply moved backwards and forwards as the wave set up travelled along the tube.

If instead of pieces of cork small balls made of beeswax (and loaded with iron-filings until they just float) are used, these when placed at various depths, and waves set up as before, will be found to move in closed curves. As the direction of motion of these vibrating particles is perpendicular to the direction of the motion of the wave, they are called **transverse vibrations**. When the direction of motion of the vibrating particles is in the direction of the wave motion, the vibrations are said to be **longitudinal vibrations**; as, for example, in the case of a sound wave in air, the wave is propagated as a series of condensations and rarefactions. As any particle of air vibrates to and fro in the direction of propagation of the wave, it is called a *longitudinal vibration*. The sound given out by a stretched wire or string, due to its vibrations when pulled on one side and let go (or bowed as in a violin string), is due to the *transverse vibration* of the wire or string.

APPARATUS.—One or two rods of mahogany, 5 or 6 ft. long, ½-inch diameter; also a glass tube about 1-inch bore, steel wire about 2 mm. diameter, V-shaped blocks of wood.

EXPERIMENT 68.—Clamp a wooden rod at its middle, and stroke one end with a resined cloth; the rod will be found to vibrate longitudinally, and the sound or note given out is due to longitudinal vibrations.

EXPERIMENT 69.—Thoroughly clean and dry the glass tube, incline it, and put in the tube a sufficient amount of lycopodium powder to form a thin covering along its internal surface. Fix at one end of the tube a well-fitting cork. The glass tube is supported in V-shaped pieces of wood. The rod used in the last experiment is firmly clamped in a V-shaped piece of wood at the middle of the rod, the clamp being fastened firmly to the table on which the apparatus rests. One end of the rod, provided with a cardboard disc of such a size as to slip easily into the glass tube, is made to enter the other end of the tube, and the supports are carefully arranged so that the axes of the rod and the tube are in the same straight line.

Set the rod vibrating by rubbing one end with a damp or resined cloth. Shift the tube, and repeat the experiment until the lycopodium is gathered into heaps at various points along the tube. The column of air enclosed in the tube is caused to vibrate by the to-and-fro movements of the vibrating rod, and the places in the tube where the lycopodium is gathered into heaps are the loops of the vibrations set up in the enclosed air. Measure the distance between as many loops as possible, and dividing by the number of loops, obtain as accurately as you can the distance between two loops; the length so obtained is half the wave length in air of the note given by the rod. If any difficulty is experienced in keeping the rod and tube in position during the rubbing, the rod may be clamped at two points, each a quarter the length of the rod from each end (instead of at the centre), and the rubbing can then take place in the middle of the rod.

SOUND. 119

Wave of Compression.

EXPERIMENT 70.—Coil a long steel wire of about 2 mm. diameter into a helix, the diameter of the coils about 7 cm. or 4 inches, and the length about 2 metres or 6 feet, the coils being about 8 mm. or $\frac{1}{3}$ inch apart. Suspend the coil from two parallel wooden rods in such a manner that the whole helix is horizontal and parallel to the rods. Gather a few coils at one end, pressing them inward, then suddenly release them. On account of the elasticity of the material, they will go back to their original position, and also beyond it. To do this a *wave of compression* is sent along the helix, each coil compressing its neighbour; so that although each coil only moves a short distance along the tube, the wave travels along the whole length. If the outer ring of the helix be pulled outwards and then let go with a jerk, a wave of rarefaction is produced, and the motion of any separate coil can be examined by attaching strips of paper or bits of twine to them.

APPARATUS.—Spring balance; one or two small weights; about a dozen marbles; board, with a groove on its upper surface; also apparatus used in Experiment 66.

EXPERIMENT 71.—If a small weight is fastened to the end of a spiral spring, then pulled down a short distance and let go, the spring will not only go back to its original position, but will go beyond it; it will then begin to return, and so will be shortened and lengthened many times before it is brought to rest.

EXPERIMENT 72.—If a dozen or more marbles be arranged in a groove (Fig. 94) and be hit with one ball, the ball at the other end will be forced away. If the row be struck by two or three balls, a corresponding number will be forced away.

EXPERIMENT 73.—If one of the marbles in the last experiment be dropped on to a smooth slab of marble or iron which is smeared with a thin coating of red ochre or lamp black, a circular smear or patch will be found on it. If the marble had simply been laid on the slab, a tiny speck showing point of contact would be observed. It is evident from this experiment that the surface of the marble was compressed, and just at the instant when brought to rest by the slab, the area of the large smear was the area in contact with the slab.

Wave of Rarefaction.

EXPERIMENT 74.—Fasten the end of the tube as in Experiment 66, and near the top fasten a piece of paper. By quickly pulling the free end a wave of rarefaction is sent up the tube; the piece of paper will be seen to move a short distance in the direction of the pull, and then to return—that is, as the tube is elastic, each small piece of the tube exerts a pull on the piece next to it, and in this manner each small piece moves a short distance along the tube and back again, thus causing what is called a *wave of rarefaction*.

Frequency is the term used to denote the number of vibrations per second. The time between any two successive passages of the vibrating particle through any point in the same direction is the **period**.

Amplitude.—When a wave of compression travels through a body, each particle of the body moves a short distance in a forward direction and then returns to its position of rest; so that if a wave of compression travels through air, each particle is made to move *forwards* a short distance and then return to its position of rest. In a similar manner a *wave of rarefaction* will cause each particle to move a little distance *backwards* and then come back to its position of rest. Thus, when a sound wave travels through air, a particle of air will move alternately forwards and backwards from its mean position. As will be seen in Experiment 76, when the spring is pulled away from the vertical and let go, the air in front is compressed; but as the air is elastic, the result corresponds to the compression of a few coils of the spring in Experiment 70: as the spring returns, the air in front expands (or is rarefied), and the state of things is analogous to that of Experiment 70. If the rod is vibrating backwards and forwards at a uniform rate, the *compressions* and *rarefactions* will follow each other *regularly*. A succession of *compressions* and *rarefactions* constitutes a **sound wave**; such a wave striking upon the ear gives rise in the brain to the sensation called sound, and requires air to carry it. The **transmission of sound** by the air particles is shown by the experiment of suspending a bell or small alarm-clock on the plate of an air-pump, and covering with a bell-shaped receiver. The sound of the alarm, distinct at first, becomes fainter and fainter as the exhaustion proceeds and the air becomes more and more rarefied. With a good pump the sound finally ceases, but again becomes distinct as air is admitted. The following simple experiment may be used for the same purpose.

Apparatus.—Glass flask with tightly-fitting cork; small toy bell fastened to cork and suspended within the flask; a short piece of glass tubing fixed in cork to allow escape of steam, the

SOUND.   121

open end of which may be closed by a piece of rubber tubing; Bunsen burner.

EXPERIMENT 75.—Boil the water to expel the air. After boiling, the flask should be closed and cooled. Note that when the bell is agitated, the sound, distinct when air was present, becomes inaudible when the air is expelled, but again becomes distinct when air is admitted.

The extreme distance through which a particle moves on each side of its mean position is called the *amplitude of the vibration*, and thus corresponds to the case of the simple pendulum.

The **wave length** is the distance which the energy travels in a medium while the body which starts it makes one complete vibration, or is the distance from any particle to the next one which is in the same phase—that is, the distance from any particle to the next particle which is moving in exactly the same manner. *The velocity with which the waves travel forward divided by the frequency gives the wave length.*

∴ If v denote the velocity of the wave, n *the frequency or number of vibrations per second*, then *wave length* $= \dfrac{v}{n}$, a formula which is true for all kinds of waves.

Vibrations of a Thin Lath.

APPARATUS.—Thin lath, or length of clock spring; vice.

EXPERIMENT 76.—Fix a long thin spring in a vice (Fig. 99). When pulled on one side of its mean position and suddenly let go, it will vibrate as shown by the dotted lines. By means of a stop watch or other convenient method of measuring time, determine the time of a vibration. Owing to frictional resistance, the amplitudes of the vibrations will gradually get less and less, but it will be found that the time taken for each vibration will be the same. Hence **the time is independent of the amplitude of the motion.**

Alter the length and repeat the experiment; it will be found that the **time of a vibration varies as the square root of the length.**

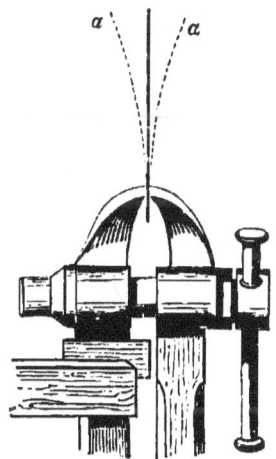

FIG. 99.—Vibration of a lath.

Denoting the length by l, tabulate as follows:—

Length.	Number of Vibrations.	t, or Time of One Vibration.	$\frac{l}{t^2}$

As in the simple pendulum, it will be found that the numbers in the last column are approximately equal.

A more satisfactory method to show that the time is independent of amplitude would be to arrange a cylinder or a plate moving at a *uniform speed*, and to allow a small style projecting from the vibrating spring to touch it. The curved line $a\,b$ (Fig. 100) represents a portion of a wave line such as would be formed by a vibrating wire or string. The dotted line AC is the axis of the waves, the arrow showing the direction in which they are travelling; $a\,e$ is the length of a wave, the length $a\,c$ being half a wave length—that is, the length moved by the wave while the particle travels from b' to a corresponding position below the line and back again to b'.

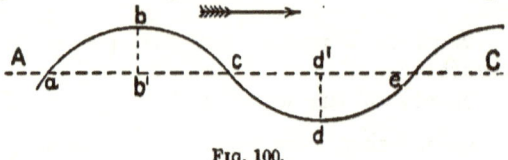

Fig. 100.

The *amplitude* is the extreme distance moved by the particle on each side of its mean position, and is represented by the distances bb' or dd', etc.

The **displacement, velocity,** or **acceleration** of any particle at any instant can be easily set out in a diagram, as shown in Fig. 97.

Thus, when any particle is at maximum displacement in positive direction, as shown by curve II., its velocity at the same instant is zero, and its acceleration is a maximum in negative direction. When displacement is zero, as at f, velocity is a maximum, as shown by the ordinate bb', and the acceleration, as shown at g, is zero. And in a similar manner the velocity, displacement, and acceleration of the vibrating point can be obtained by comparing the sinous curves I., II., and III. (Fig. 101).

Instead of the spring used in Experiment 76, we may use the following:—

APPARATUS.—A piece of clock spring about one foot in length, one end filed to a point, to which a silvered bead is fastened. The middle part of the spring is softened by the flame of a Bunsen burner and twisted sharply, so that the planes of the two ends are at right angles to each other.

EXPERIMENT 77.— Fasten the lower end in the vice; then when pulled on one side and let go, a curve is traced by the bead on the free end of the spring.

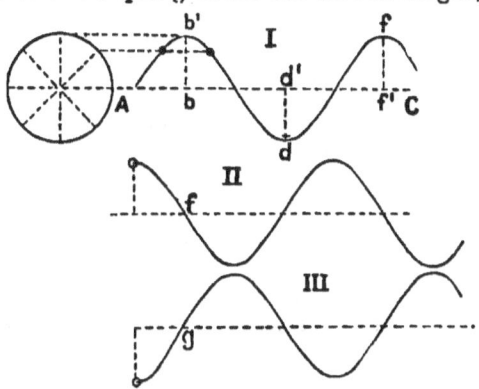

FIG. 101.—Curves of velocity, displacement, and acceleration.

SUMMARY.

Sound Waves are caused by the vibrations of the particles of air giving rise to a series of alternate *compressions* and *rarefactions*.

Transverse vibrations are vibrations at right angles to the length of a body.
Longitudinal vibrations are vibrations in the direction of the length of a body.
Frequency denotes the number of vibrations per second.
Amplitude denotes the extreme distance in either direction through which any particle moves from its mean position.
Wave Length is the distance from any particle to the next one which is moving in exactly the same manner as the first; or, velocity ÷ frequency.

EXERCISES.

1. Waves of sound the frequency of which is 256 vibrations per second pass from a stratum of cold air to a layer of hot air. In the cold air the velocity is 1,120, and in the hot air 1,132 feet per second. Find the wave length in each case.
2. Explain how the condensation and rarefaction constituting a wave of sound are produced. How is a sound wave propagated through air? What is meant by a *wave of sound* and by the length of a wave?
3. A glass rod 5 feet long is clamped at its centre and rubbed longitudinally with a wet cloth. State how it vibrates when thus treated, and calculate the velocity of sound in glass, if the above rod makes 1,295 complete vibrations every second.
4. What do you understand by a vibratory motion? Distinguish between longitudinal and transverse vibrations, and illustrate your answer by an example of each.
5. Does the time of oscillation of a pendulum depend on the amplitude of the vibration? What bearing has your answer on the theory of sound?
6. Define the meaning of the term "frequency." If the frequency of a tuning-fork is 128, and the number of vibrations per hour of a second fork exceeds that of the first by 300, how many beats will there be in a minute if the two are sounded together?

CHAPTER X.

VIBRATIONS OF A STRETCHED WIRE OR STRING MONOCHORD.

Transverse Vibrations.—To show that the **frequency or number of vibrations per second** made by a stretched string or wire depends upon—

(1) *Material*, (2) *stretching weight or tension*, (3) *thickness of the string or wire*, (4) *length*.

An instrument for the study of the vibrations of stretched wires or string is called a *monochord* or sonometer. Or the top of a pine table on which two blocks are fixed may be used. Over these blocks the string or wire is stretched. A weight may be used to stretch the wire, or the tension may be indicated by using a spring balance.

Monochord.—A deal board 3 feet long, 6 inches wide, and 1 inch thick, having one end bevelled, into which two round-headed screws are fixed in such a manner that they slant outwards. Two hard-wood bridges are fixed, one at each end; these may be made 5 inches long and 1 inch square, having the upper edges bevelled, and at the centre a piece of thick brass wire inserted, on which the steel wires can rest. Two hard-wood bridges made of the same dimensions and in a similar manner are used as movable bridges. A piece of hard wood is firmly fixed into the board at one end, so that a wrest-pin can be carried which fits tightly and allows the wire to which it is attached to be tightened by turning the pin; this wire is used as a standard of reference, which can be done by using a tuning-fork and turning the pin until the two are in unison. A wire fastened to the other peg carries at its lower end a hook or scale pan to which weights can be attached; the board is fixed in a nearly vertical position, having the lower end sloping outwards, so that the wires bear on the lower bridges.

VIBRATIONS OF A STRETCHED WIRE.   125

A scale is fixed to the board, accurately graduated either into inches and fractions of an inch, or into centimetres and millimetres.

APPARATUS.—Weights; steel and brass pianoforte wire of different thickness, and monochord.

Number of Vibrations Varies inversely as the Length.

EXPERIMENT 78.—Twist loops on the ends of two pianoforte wires (of same diameter and material); put them over the two pegs; fasten one to the wrest-pin, and to the other attach any suitable weight. The bridges being at equal distances apart, adjust the tension, or pull in one wire by turning the wrest-pin until, when the strings are plucked, the same note is obtained from each—that is, the wires are in unison. When this is obtained, move the bridge under the wire to which the weight is attached nearer to the fixed end. It will be found that as the length of the wire (which is the distance between the bridges) is *shortened* the sound becomes higher in pitch, the cause being the *increased number of vibrations*.

The converse is also seen to be true—that is, if the length be increased, the number of vibrations is diminished.

This is also to be observed in Experiment 76, the strip being fastened in the vice, noting the number of vibrations made by a projecting length, and the increased number when the projecting part is shortened.

When one bridge is half-way along the board, the length of one string being half that of the other, the number of vibrations made by the shorter one will be found to be twice that of the longer string, and the shorter will give the octave of the note given by the longer string.

Thus, when the two strings of the same material and same diameter are stretched by equal weights, the number of vibrations, or *the pitch*, varies inversely as the length.

Musical Sounds or Notes.—When a sound is such that its effect on the ear is continuous and agreeable, it is called a musical sound. A musical sound is produced by a number of vibrations reaching the ear with regularity, and in such rapid succession that no interval of time can be detected between them; in other words, the sound is continuous. These sounds differ in *pitch*. When the number of vibrations per second is great, the pitch is high; when small, the pitch is low.

The **pitch of a note** may be expressed either relatively with regard to some other note adopted as a standard, or absolutely by the number of vibrations required to produce the note. Thus the middle C corresponds to 256 vibrations per second, and its octave to 2 × 256, or 512 per second.

In **Savart's wheel** a toothed wheel is fitted to a whirling table. As the wheel rotates the teeth are made to touch lightly a thin card. When the wheel is moving slowly a series of regular taps is heard. These occur much more rapidly as the speed of the wheel is increased, until they blend together, producing a low note, the pitch of which rises as the speed is increased.

Disc Siren.—A jet of air is blown against a rotating disc in which holes have been pierced. The holes are arranged in concentric circles. Then, as the disc is rotated, each time a hole comes opposite the jet, the air rushes through the opening, producing a wave of compression. Owing to the elasticity of the air, the wave of compression is succeeded by a wave of rarefaction during the short interval that elapses until the next hole comes into position. The interval is diminished as the speed of the wheel is increased; the pitch of the note produced rises with the speed of the disc.

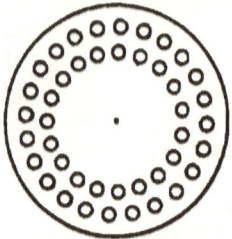

Fig. 102.—Disc siren.

The monochord may be used to exhibit the relations which exist between the notes of the musical intervals, by altering the length of the string.

EXPERIMENT 79.—Divide the distance between the two fixed bridges (as indicated by the scale underneath the wire) into the reciprocals of the following fractions, then sound the string as a whole, and afterwards shift the movable bridge; the diatonic scale will be produced—

C	D	E	F	G	A	B	Oct. of C.
1	$\frac{9}{8}$	$\frac{5}{4}$	$\frac{4}{3}$	$\frac{3}{2}$	$\frac{5}{3}$	$\frac{15}{8}$	$\frac{2}{1}$

The disc siren may be used to explain the harmonious intervals, such as, for example, the common or major chord C E G C. The disc should be perforated with four concentric rows of holes; the numbers of holes in the rows, beginning with the outer, should be as 8, 6, 5, and 4.

If in Experiment 79 the wire be plucked gently, so that the amplitude of the vibration is small, the sound given out by the vibrating string will be almost inaudible; but if the amplitude be increased, the loudness of the note given out will be found to increase with the amplitude, or more exactly, *the loudness is proportional to the square of the amplitude.*

Pitch Varies as the Square Root of the Tension.

EXPERIMENT 80.—Place the two bridges so that the distances between them, or the length of the wires, is the same. To one wire attach any suitable weight, and adjust the wrest-pin until the two wires when set vibrating are in unison. If now the weight be increased, the number of vibrations will be found to increase, and the pitch is raised; when the weight attached becomes four times the preceding amount, the number of vibrations will be doubled —in other words, one wire will give a note and the other its octave. Hence *the pitch varies as the square root of the tension.*

If the wrest-pin be so adjusted that one wire gives when plucked the note C (this can be tested by using a tuning-fork vibrating 256 times per second), and if the other wire be weighted until it gives the note C, and the weight increased until the note E is obtained, it will be found that the two loads are as 16 to 25, the square roots of which are 4 and 5. Hence the number of vibrations in the heavier-weighted string are $\frac{5}{4}$ those of the other, or $\frac{5}{4} \times 256 = 320$ vibrations per second.

In a similar manner, to produce the note G, or 384 vibrations per second, the weights are as 9 to 4.

EXPERIMENT 81.—Weigh equal lengths of brass and steel wire; attach to the monochord, and load with equal weights: the number of vibrations will be found to be different. Alter the lengths of the wires until unison between the notes given out by the wires is obtained; it will be found that the lengths are in *inverse proportion to the square roots of the weights of equal lengths of the wires*, or that with equal lengths of wires of different density the number of vibrations is inversely proportional to the square roots of the weights of equal lengths of the wires.

A violin furnishes a good illustration. The lowest or bass string is wrapped round with metal wire; the tension is adjusted by means of tightening-pegs. The three upper strings are of different thicknesses, and the length is adjusted by the "fingering," shortening or lengthening the vibrating portion so as to obtain different notes from the same string.

All the preceding may be expressed in the following formula:—

$$N = \tfrac{1}{2} \sqrt{\frac{F}{lm}} \quad \dots \dots \dots \dots (1),$$

where N = number of vibrations per second, or frequency; l the length, and m the mass of the wire.

If ρ denote the density of the wire, then

$$m = \frac{\pi}{4} d^2 l \rho.$$

Hence substituting in Eq. (1), we get

$$N = \frac{1}{dl} \sqrt{\frac{F}{\pi \rho}}.$$

Resonance.

APPARATUS.—Tuning-fork; glass tube; cork, fitted so as to slide easily in and out of the tube, fastened to a piece of brass wire. (Fig. 103.)

EXPERIMENT 82.—Hold the prongs of the vibrating fork with its plane vertical, as shown in Fig. 103. Shorten the air column by pushing in the cork: the sound becomes louder until a certain point is reached; beyond this the sound gets fainter as the column shortens. Suppose the dotted lines to represent one swing of the fork from a to b. From a to b a pulse of condensation runs down the tube, is reflected, and returns to the mouth. When it swings back from b to a (if the length A B is just the right length), the condensation will travel to the bottom and back just in time to combine with the condensation produced outside as it moves from b to a. In a similar manner a rarefaction goes to the bottom and back to A as the prong travels from b to a. When the air column is longer or shorter, the waves do not reach A at the right moment. In the time required for a complete vibration, a pulse of condensation and a pulse of rarefaction have travelled each twice the length of the tube. If the tube had been open (the cork removed), the wave would have travelled a distance equal to four times the length of tube A B.

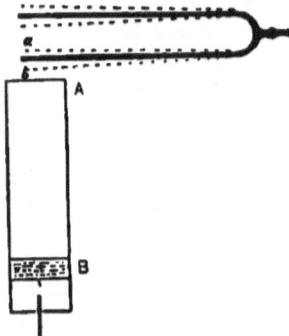

FIG. 103.—Resonance of air column.

SUMMARY.

Transverse Vibrations.—Frequency (or number of vibrations per second) of a stretched wire or string depends upon (1) material, (2) tension, (3) thickness or diameter, (4) length.

Pitch depends upon frequency. The pitch of a note is raised an octave when the frequency is doubled.

Exercises.

1. Explain the character of the vibration of a stretched violin string. What effect is produced by touching it at one-third of its length from end to end?

2. What is the relation between the wave length and the frequency which corresponds to a given musical note? On what experiment is your answer based?

3. Two trains, moving equally fast, approach and pass each other. An observer in one of them notices that the note emitted by the whistle of the engine of the other train changes, as it passes him, from 512 to 480 vibrations per second. At what rate do the trains pass each other, and what is the true vibration frequency of the note of the whistle? ($N.B.$—Take velocity of sound = 1,100 feet per second.)

4. State in what way the rate of transverse vibration of a stretched string depends upon the tension. How would you determine the rate of vibration of the string?

5. A steel wire, one yard long, and stretched by a weight of 5 lbs., vibrates 100 times per second when plucked. If I wish to make two yards of the same wire vibrate *twice* as fast, with what weight must I stretch it?

6. If two equally-stretched strings of the same thickness, one of steel, the other of catgut, give the same note when struck, which of them is the longer?

7. A cog-wheel containing 64 cogs revolves 240 times per minute. What is the frequency of the musical note produced when a card is held against the revolving teeth? Find also the wave length corresponding to the note if the velocity of sound is 1126·4 feet per second.

8. A wheel with 33 teeth touches a card as it spins, and thereby emits a note two octaves above the middle C, which has 256 vibrations per second. How many revolutions is the wheel making per minute?

9. A circular disc pierced with a concentric ring of equidistant holes rotates about an axle which passes through its centre. A stream of air from a small jet is blown against the ring of holes. On what do the intensity and the pitch of the note produced depend?

CHAPTER XI.

LIGHT.

Light Waves.—In the preceding chapters on **sound**, we have considered sound waves. A sounding body acting on the neighbouring air particles sets them oscillating backwards and forwards; these in turn act on others all along the line of travel of the sound wave. Such an oscillation of the particles to and fro, in which, after the disturbance, the air particles return to their normal condition, is called a *wave-motion*. We next consider another kind of wave-motion, and to understand the manner in which light is produced, we assume the waves to be transmitted by something in space which we cannot see or otherwise detect. The medium which is assumed by physicists to exist is called **ether**, and is supposed to fill all space, entering into the mass of all bodies, and filling the little spaces between the minute particles of which all bodies are supposed to consist. The energy residing in **ether waves**, coming to our planet from the great source of energy the sun, is called **radiant energy**, and receives three different names. Thus, if the ether waves raise the temperature of the receiving body, it is called **radiant heat**; if they affect the eye, they produce **light**; when they effect chemical changes, as on a photographic plate, it is called **actinic energy**.

Light Travels in Straight Lines.—That light travels in straight lines can easily be verified. Thus a beam of light from a lamp passing through a small round hole in the window-shutter of a darkened room forms a divergent or conical beam. If a sheet of paper be held in the path of the beam and at right angles to it, a round, bright patch of light is formed, the size of which increases as the distance of the paper from the shutter is increased.

APPARATUS.—Cardboard, candle, tissue paper or tracing paper.

LIGHT.

EXPERIMENT 83.—In a sheet of cardboard make a small pinhole, as shown at o (Fig. 104). Place at S a screen made of tissue

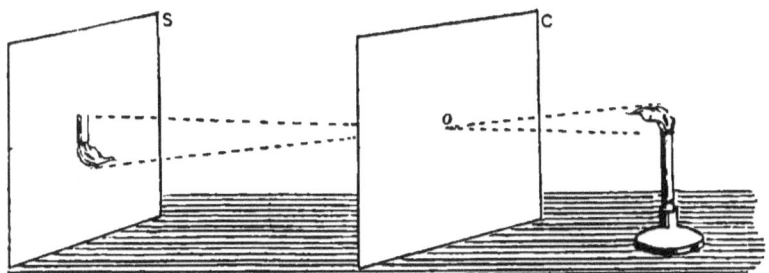

FIG. 104.—Experiment to show that light travels in straight lines.

or tracing paper, and a lighted candle in such a position that the centre of the candle flame is at or about the height of the pinhole o.

When this is done, an inverted image is seen on the screen.

If the hole o were enlarged, instead of a clearly-defined image, a blurred image would be formed on the screen.

If the screen S be removed, and replaced by a sheet of cardboard having a pinhole as shown at o, then when placed so that the two holes are at the same height and in a straight line, the light can be seen shining through, but is cut off as soon as either card is moved.

Law of Inverse Squares.

APPARATUS.—A small, square wire frame; candle.

EXPERIMENT 84.—Fix the wire frame A B (Fig. 105) in a vertical position, between the lighted candle and a screen, or the

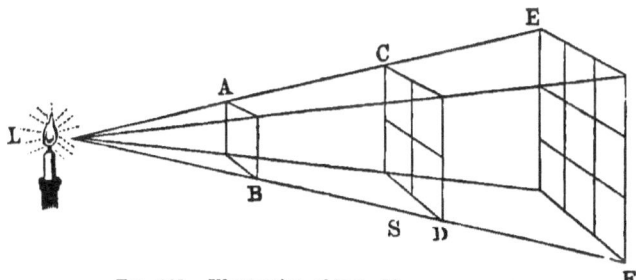

FIG. 105.—Illustration of law of inverse squares.

wall of a darkened room, the flame of the candle being at the same height as the centre of the frame. Cut out a piece of paper which just fits the shadow, C D, cast on the screen S, or the wall. If the paper be folded twice across, as shown by the lines, the size of each of the four squares into which S is divided will be found to equal the size of the wire frame. Hence the area of the square S is four times that of the frame. As the same amount of light which is cast upon A B is cast upon the square S, the intensity of illumination on S is only $\frac{1}{4}$ that on A B. If the distances are 1 and 3, the areas will be 1 to 9, as shown at E F.

Repeat the experiment, altering the distance of A B from L. Tabulate as follows:—

Distance of A B.	Distance of S.	Area A B.	Area of S.

Rumford's or Shadow Photometer.

APPARATUS.—A cardboard or ground-glass screen, fixed in a vertical position; a small rod 12 or 14 inches in length, and $\frac{1}{2}$ inch diameter, fixed in an upright position opposite the centre of the screen, and about 6 inches from it; candle, and gas or lamp.

EXPERIMENT 85.—Arrange the flame of the gas (or the lamp) almost edgewise, and about 3 or 4 feet from the screen, so that the flame of the gas and that of the candle are about equal in height. Adjust the position of the candle until the two shadows cast by the rod on the white screen lie edge to edge, but do not overlap (Fig. 106).

FIG. 106.—Shadow photometer.

When the shadows are equally dark, the screen is equally illuminated by the two lights. Measure the distances from the screen

LIGHT. 133

to the candle and to the gas flame. Then the squares of these distances will represent the comparative illuminating powers of the two lights.

Repeat the experiment with the gas flame at different distances from the screen, in each case measuring carefully the distance of the candle and gas or lamp flames.

Tabulate as follows :—

Distance of candle from screen $= c$	Distance of gas or lamp from screen $= l$.	l^2.	c^2.	$\dfrac{l^2}{c^2}$.

It will be found that the numbers in the last column are either equal or very nearly so. From this result the illuminating (or light-giving) powers of two sources are to each other *as the squares of their respective distances from any surface which they illuminate equally.*

EXAMPLE.—If the lamp at 5 feet from the screen throws a shadow as dark as a candle at 3 feet from it, the illuminating powers of the two lights are as 5^2 to 3^2, or as 25 to 9, and thus at any distance l from the screen, the candle being at a distance c, the comparative illuminating powers of the two lights are as l^2 to c^2.

The unit used to measure the light-giving power of various sources of light is the light-giving power of a sperm candle (6 to the lb., and burning at the rate of 120 grains per hour).

Grease Spot or Bunsen's Photometer.

When a grease spot is made on a sheet of unglazed paper (which is easily effected by dropping some melted wax from a paraffin candle), and a light placed on one side the sheet, it will be found that the light is transmitted quite freely through the spot.

APPARATUS.—A sheet of cardboard about $8'' \times 4''$ through which a hole is cut about 2 inches square; a piece of white unglazed paper is pasted over the hole, at the centre of which a grease spot is made, and the card fixed in a vertical position.

EXPERIMENT 86.—Place a lighted candle on one side, and the spot will appear darker than the rest of the sheet when looked at from the side on which the candle is placed, but much lighter when looked at from the other side. If now a lighted candle be placed on the other side, and the candle moved about until the grease spot and the rest of the sheet are equally bright, so that the spot becomes either altogether or nearly invisible, the two candles (giving about the same amount of light) will be found to be at equal distances from the sheet. If the candle on one side the sheet be moved so that its distance from the screen is twice what it was before, it will be found that three additional candles are required, or four candles on one side at double the distance of one candle on the other side, to give the same *intensity of light*—thus proving that *the intensity of light varies inversely as the square of the distance.*

Hence, to compare the illuminating power of two sources of light, they are placed one on each side of the screen, and on a level with the grease spot. The two sources are so adjusted that the grease spot is of the same brightness as the rest of the screen, and the distances are measured; then *the illuminating power of the two sources are as the squares of their respective distances from the screen.*

EXAMPLE.—Place a gas flame and a candle flame at the same height, and about 5 or 6 feet apart. Move the cardboard sheet between the two flames until both sides of the sheet become equally illuminated and the bright grease spot disappears, or nearly so. Measure the distances of the two flames from the screen. If these are 4 feet and 1 foot respectively, the intensities of the lights are proportional to 16 and 1, or 4^2 and 1^2. If a standard candle were used in the experiment, the gas flame would be said to be 16 candle-power. In a similar manner the intensity of any other flame may be found.

A convenient form of photometer was introduced by Bunsen, simple in construction, and similar in principle to the above. It is much used to test the illuminating power of coal-gas, oil-lamps, and other flames.

SUMMARY.

Light, sound, and every kind of radiation is called *radiant energy*. In the transference of energy by the ether waves, if the temperature of the receiving body be increased, it is called *radiant heat;* if the waves affect the retina, they are known as light.

Propagation of Light.—Light travels in straight lines when passing through one medium, but is usually refracted—that is, has its direction changed—in passing from one medium to another.

LIGHT.

Illumination of a body is the amount of light received from a given source, and is found to *vary inversely as the square of the distance of the body from the source of light.*

Photometer is an instrument used for comparing the intensities of two sources of light.

Rumford's or Shadow Photometer.—The light-giving powers of two sources are as the *squares of their distances from the screen.*

The Grease Spot Photometer.—The intensities of the two sources of light are as the *squares of their distances from the spot.*

Exercises.

1. What do you understand by the intensity of the illumination at a point due to a given source? Describe experiments to prove that the intensity of the illumination at a point due to a given small source is inversely proportional to the square of the distance of the point from the source.

2. Two sources of light when placed at 8 and 10 feet from a screen produce the same intensity of illumination. Compare the illuminating powers of the two sources.

3. If the intensity of illumination of a screen at 4 feet from a given source of light be denoted by 1, find the intensity when the distance of the screen is increased to 9 feet.

4. Describe a simple photometer, and state how it is used for comparing the intensities of two sources of light.

5. Of two gas flames, one gives out twenty-five times as much light as the other. If the smaller flame be at a distance of 2 feet from the screen, what is the distance of the larger flame when the intensity of light is equal on both sides?

6. What course would you adopt if it were necessary to compare the illuminating power of a very bright light with that of a standard candle?

7. Explain why the clear image of a brightly-illuminated object, which can be formed on a screen by means of a pin-hole, becomes blurred if the hole is enlarged. Illustrate your answer by a diagram.

CHAPTER XII.

REFLECTION—REFRACTION.

Light is reflected either irregularly or regularly—irregularly from the surface of rough or imperfectly-polished bodies, and regularly from smooth, polished surfaces.

When a ray of light falls upon a polished surface, it is reflected or thrown back from that surface. If the light be perpendicular to the surface, it is reflected back along the path by which it came; but when the ray is oblique to the surface, the reflection is also oblique. Also the normal to the surface at the point of incidence, and the incident and reflected rays, are in the same plane. These facts are expressed in the following two laws, called the **Laws of Reflection**:—1. The angle made by the incident ray with the normal is always equal to the angle which the reflected ray makes with the normal, or **the angle of incidence is equal to the angle of reflection.** 2. The **incident ray**, the **reflected ray**, and the **normal**, at the point of incidence, **all lie in one plane.**

APPARATUS.—Small wooden stick (whitened) fastened with wax at the centre of a flat mirror and perpendicular to the surface.

EXPERIMENT 87.—If a beam of parallel rays from a lantern, or a sunbeam passing through a hole in a screen, be cast upon the mirror at the foot of the rod, it will be found that the angles which the reflected and incident beams make with the mirror and stick are equal to each other. Thus in Fig. 107, A B is the mirror, bS the stick perpendicular to A B, cb and bd the incident and the reflected beams respectively: the angle made by the reflected beam with the mirror is the angle

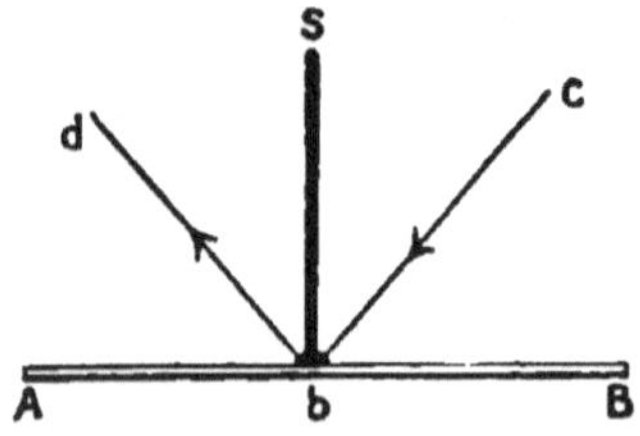

FIG. 107.—Angle of incidence is equal to angle of reflection.

Abd, and this is equal to the angle cbB, which the incident beam cb makes with the mirror. Also the angles dbS and Sbc made by the reflected and incident beams respectively with the stick are equal. The stick represents the normal to the mirror. Hence the **angle of incidence is equal to the angle of reflection.**

By using a sheet of stiff paper or cardboard and placing it in position, it will be found that *the incident beam, the normal* (or stick), *and the reflected beam lie in one plane.*

Also, if the mirror be moved through any angle, the reflected beam moves through an angle twice as great : thus, if the mirror were held so that bc coincided with bS, the angle of incidence would be 0°, and therefore the angle of reflection would also be 0°. If now the mirror be turned through 15°, the angle Sbc would be 15°, and Sbd would be 15°; but as bd originally coincided with bc, by turning the mirror through 15° the line bd makes an angle of 30° with its original direction. In a similar manner for any other angle the rule holds.

To verify the Laws of Reflection.

APPARATUS.—A flat board ABCD (Fig. 108) on which a graduated semicircle is fixed; a small piece of thin glass m. The back of the glass should be blackened so that reflection only takes place from the front; when glass silvered at the back is used, the refraction through the glass may cause considerable

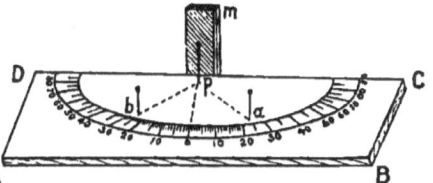

FIG. 108.—Apparatus to verify the laws of reflection.

error. Three toilet pins, 3 or 4 inches long, with black round heads.

EXPERIMENT 88.—Draw a line op passing through o, the zero of the graduated scale, and perpendicular to the edge CD (Fig. 107): this represents the normal. Fix a pin at p close to the mirror, and another in any convenient position b; place the third pin, a, in such a position that a, p, and the image of b are in the same straight line; when this is done, note that the angles bpo and opa are equal. A sheet of drawing-paper may be used instead of the graduated scale shown; the lines bp, pa, and DC are carefully drawn on the paper, and finally the line po at right angles to the edge DC.

Refraction.

EXPERIMENT 89.—Place a small disc, such as a coin, in any convenient vessel, such as a basin, and place the basin in such a

138   REFLECTION—REFRACTION.

position that when looked at from a point E the coin is just hidden by the edge of the basin (Fig. 109). If now water be poured into the basin as carefully as possible, so as not to disturb the position of the coin, the coin appears to rise with the level of the water until its whole surface is visible. Any rays, such as E I and E II, meeting the surface of the water at I and II, are reflected away from the normal, as shown by I d, II b; but to the eye at E the edges d and b appear to be at d' and b' in E I and E II produced.

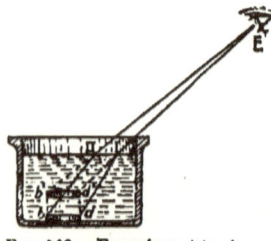

FIG. 100.—Experiment to show refraction.

Refraction.—When a ray of light falls perpendicularly upon any surface, it continues its course in the same straight line; but a ray of light which falls obliquely upon the surface of a transparent body is found to be divided into two parts, one of which enters the body, the other is reflected from the surface.

If a ray passes obliquely from one medium into another of different density, it is found to change its direction. Thus a ray of light R (Fig. 110), which reaches the surface of water at I, as shown, is deviated from the line R I, and goes on in a new direction I C. Let M M' denote the normal to the surface at I, the reflected part I R' is such that the **angle of incidence** R I M **is equal to the angle of reflection** M I R'; if the line R I be produced to any point B', then the *angle* C I M' *is called the* **angle of refraction.**
The angle B' I C shows the deviation of the ray from the straight line, and is called *the* **angle of deviation.**

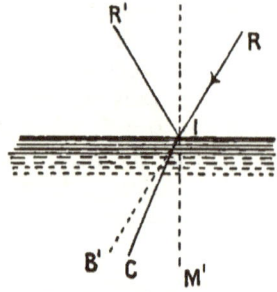

FIG. 110.—Refraction.

Thus, when a ray of light passes from a rarer to a denser medium, *the angle of refraction is always less than the angle of incidence;* when the ray passes from a denser to a rarer medium—say from water to air—the angle is greater.

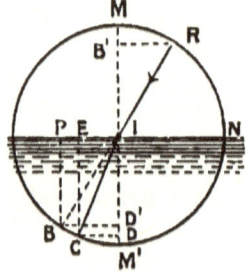

FIG. 111.—Angle of refraction.

Given the path of a ray in air, to find the path in water. If R I is the incident ray (Fig. 111) at the point of incidence (I) with any convenient

REFLECTION—REFRACTION. 139

radius, describe a circle; draw M M' perpendicular to the surface. Let R be the point where the incident ray cuts the circle; produce R I to cut the circle again at B; from B drop a perpendicular B D' on the normal, and make I P equal to B D'. Divide I P into four equal parts, making $IE = \frac{3}{4} IP$.

Draw E C parallel to M M', cutting the circle at C; join C to I, then I C is the path of the refracted ray, and C I M' **the angle of refraction.** If C D be drawn perpendicular to M M', the ratio of B D' to C D is the **index of refraction**, and is usually denoted by μ.

When the two media are air and water, $\mu = \dfrac{BD'}{CD} = \dfrac{4}{3}$.

As the object is only to obtain the direction I C, the following simple geometrical construction may be used:—

Let R I (Fig. 112) denote a ray incident to the surface of the water N N'; draw the normal M M', and R B' perpendicular to M M'; with any convenient radius describe arc of circle R B M. Make $B'P = \dfrac{3}{4} B'R$; draw P B perpendicular to R B', meeting the circular arc in B; join B to I and produce to C, then I C represents the path of the refracted ray.

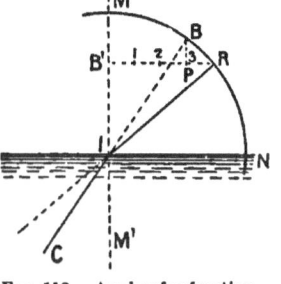

Fig. 112.—Angle of refraction.

If the second medium be glass $\left(\mu = \dfrac{3}{2}\right)$, make $B'P = \dfrac{2}{3} B'R$, and proceed as before.

When a ray passes from a denser to a rarer medium, the construction is the same as described. Thus:—

Given the path of a ray of light in water, to find the path in air. When a ray passes from water to air, $\mu = \dfrac{3}{4}$. Hence if C I (Fig. 111) denote the ray, with I as centre and any convenient radius, draw a circle; draw the normal M M' passing through I, and at the point of intersection (C) of the ray (C I) with the circle draw C E parallel to M M'. Divide I E into three equal parts, and make I P equal to four of these parts; draw P B parallel to the normal, and intersecting the circle at B. Join B I, and produce to cut the circle at R; then I R represents the path of the refracted ray in air.

In passing from the medium of greater to one of less refractive power, the ray is bent away from the normal.

Critical Angle.—As the angle C′ I M′ increases, the point R (Fig. 112) approaches the surface N; and when R coincides with N, the refracted ray just touches the surface of the water. In Fig. 113,

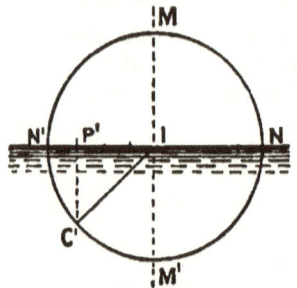

Fig. 113.—Critical angle.

make I P′ = $\frac{3}{4}$ I N′. Draw P′ C′ parallel to M M′, cutting the circle at C′; join C′ to I. Then C′ I M determines the **critical angle**—that is, for any angle greater than C′ I M the point R would fall below the surface; hence there would be no refracted ray, and the light would be totally reflected.

In water with refractive index 1·33, critical angle = 48·5°.

With crown glass refractive index 1·56, critical angle = 41°.

Refractive Index of Water.

APPARATUS.—A small rectangular trough (which may be a biscuit tin made water-tight by marine glue); a sheet of blackened paper fastened to a block of wood, and having a small hole as shown at O (Fig. 114).

EXPERIMENT 90.—Allow a beam of light (from a lantern, or by using the apparatus in a dark room) to pass through the opening O (Fig. 108). If a scale be placed along the base of the vessel, the reading of the point B can be noted; if now the water be carefully run in so that the position of the scale or the vessel is not disturbed, it will be found that the beam will be deviated from the straight line R B, and will be refracted towards the normal M M′; when the level of the water reaches a known height I, the reading of the point C is taken; then by means of a sheet of paper and drawing instruments the values can be verified.

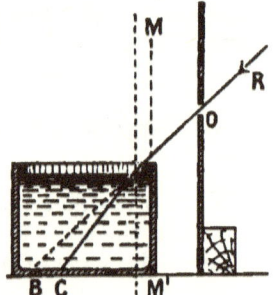

Fig. 114.—Experiment to determine angle of refraction.

The angle M′ I B is called the *angle of incidence*, also M′ I C is the *angle of refraction*, and B I C the *angle of deviation*.

Refractive Index of Glass.

APPARATUS.—A rectangular block of glass (a piece of glass from a box of weights may be used); a few black-headed toilet pins; drawing paper and instruments.

EXPERIMENT 91.—Pin a sheet of drawing-paper on a drawing-board, as shown in Fig. 115; place two pins, a and b, in any convenient position, the pin b being close to the block of glass.

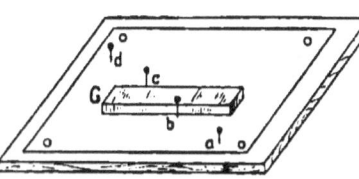

FIG. 115.—Refraction through glass.

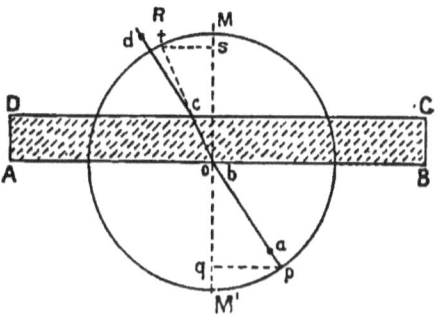

FIG. 116.—Refraction through glass.

When looking through the glass plate G from the opposite side of a and b, place the two pins c and d in such position that all four pins appear to be in the same straight line. By means of a finely-pointed pencil trace carefully the outline of the A B C D (Fig. 116) of the block of glass on the paper; next take away the glass and the pins, and join the points $a\,b\,c\,d$.

At the point o, where the line $b\,c$ cuts the edge A B, draw a circle of any convenient size, using o as centre; draw also the normal M M' passing through o.

Draw $p\,q$ and $r\,s$ perpendicular to the normal M M'. Measure the lengths of $p\,q$ and $r\,s$, and obtain the ratio of $\frac{p\,q}{r\,s}$. Alter the positions of the pins, and proceed as before; in each case obtain the ratio $\frac{p\,q}{r\,s}$. You will find that the ratio so obtained is approximately the same.

It will be found that the ray emerging from the glass at c continues along the line $c\,d$, which is parallel to $a\,b$, but not in the same straight line.

Thus a ray of light incident to a glass plate, as shown in Fig. 116, in passing through the denser medium is bent towards the normal, and emerging continues along a line parallel to the incident ray, but not in the same straight line.

Summary.

Reflection.—Light, reflected from a suitable surface, is found to obey the following, called the **Laws of Reflection**:—1. The *reflected* ray, the *normal* at the point of incidence, and the *incident* ray *all lie in one plane*. 2. The **angle of reflection is equal to the angle of incidence**.

Refraction.—A ray of light in passing from one medium to another of different density does not continue in a straight line, but is *bent*. The angle between the normal and the path of the ray in the refracting medium is called the **angle of refraction**.

When a ray of light passes from a rarer to a denser medium, such as from air to water, *the angle of refraction is always less than the angle of incidence*.

Total Reflection.—As the angle of incidence within the denser medium increases, a certain value is obtained beyond which the rays do not pass out, but are totally reflected; such an angle is called the **critical angle**.

Refraction through a Glass Plate.—When a ray of light passes through a glass plate, the light is *refracted—the emergent ray is parallel to the incident ray*.

Exercises.

1. Account for the appearance of a straight rod dipped obliquely into water, illustrating your answer by a diagram.

2. A ray of light passing from air to water falls at a given angle on the surface of the water. Give a geometrical construction to determine the path of the refracted ray, $\mu = \frac{4}{3}$.

3. You are given a rectangular block of glass, a drawing-board, and drawing materials, and some pins. How would you use these to find the refractive index of the glass?

4. The image of a distant bright point situated vertically over a vessel of water is formed on the bottom of the vessel by a convex glass lens immersed in the water. Would the distance of the lens from the bottom be the same if the vessel were filled with air? If not, explain with diagrams the cause of the difference.

CHAPTER XIII.

MIRRORS—LENSES.

Inclined Mirrors.

APPARATUS.—Two small plane mirrors about 8 inches by 6 inches, having two edges hinged together, which can be effected by pasting on a strip of cloth; candle, and a semicircle of cardboard divided into degrees.

EXPERIMENT 92.—Place the two mirrors in a vertical position on a table with the lighted candle between them, and a number of images of the candle flame are seen; it will be found that the number of images depends on the angle at which the two mirrors are inclined. Thus, if the mirrors are at right angles, or the angle between them is 90°, the number of images will be three; if the angle be 60°, the number will be 5; if 45°, the number is 7; or if $\theta°$ be the angle between the mirrors, then the number of images is given by $\dfrac{360°}{\theta} - 1$.

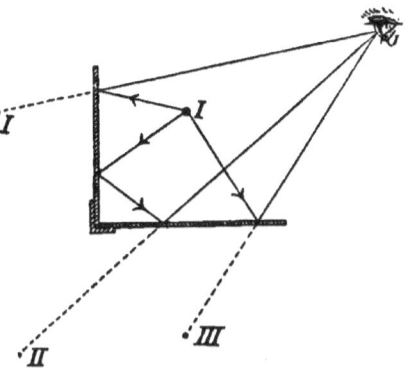

FIG. 117.—Experiment with two mirrors at right angles.

With two plane mirrors at right angles, let I be the luminous point (Fig. 117). Three virtual images are seen by the eye—one at I, another at II, and the third at III.

Multiple Images.—A ray of light from a point R falls upon a plate glass mirror GG (Fig. 118). From the front surface of the glass the image R_1 is formed; this, when R is nearly at right angles to the glass, is a faint one. The second image, R_0, is caused by reflection from the silvered back, and is brightest. The light is reflected backwards and forwards as shown in Fig. 117, producing a series of images, R_2, R_3, etc., of gradually-decreasing brightness.

APPARATUS.—Glass prism; pins; drawing-paper.

EXPERIMENT 93.—Fix a piece of drawing-paper on a drawing-board, or on the top of a table; on the paper place the glass prism in an upright position (Fig. 119). Stick two pins into the paper, as shown at R and A; on the other side of the prism stick two pins at A' and R_1, in such positions that, on looking through the prism, the four pins appear to be in the same straight line; next trace the outline of the prism on the paper by means of a fine-pointed pencil, removing the pins and the prism; join the centres of the pin-holes by straight lines, so obtaining the lines RA, A'A', and A'R_1. As the ray RA passes from a rarer to a denser medium, the ray is bent towards the normal, and therefore towards the base of the prism, as shown by AA'. On emerging at A' it is bent away from the normal to FD; hence it passes in a direction A'R'.

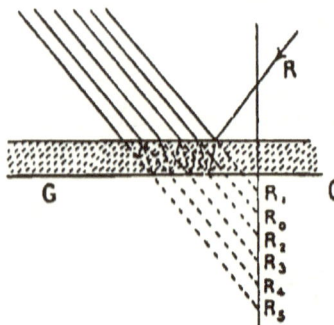

FIG. 118.—Multiple Images.

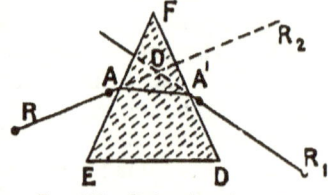

FIG. 119.—Refraction through a glass prism.

The angle $R_2 O R_1$ is called the *angle of deviation*.

If a second wedge, similar to the first, is placed so that the two form a single wedge of double the thickness, the beam will be displaced through twice the distance (that is, if the ray emerges and is not totally reflected; this, of course, depends on the inclination of the incident ray). If the two wedges be now placed so that a rectangular mass of glass of same thickness throughout is formed, as before it will be found that the emergent and incident beams will be parallel, but not in the same straight line (the two prisms acting as a parallel plate of glass).

If the sides or faces of the wedge-shaped prism (Fig. 119) were curved and convex, then two such wedges placed base to base would give what is called a convex lens. **Lenses** may be divided

MIRRORS—LENSES.

into two classes: those *thicker at the centre than at the edges* are called **convex** or **converging lenses**, and those *thicker at the edges than at the centre* are called **concave** or **diverging**. Sections of six forms of lenses are shown in Fig. 120. At A is shown a section of a *double-convex lens*, having two convex sur-

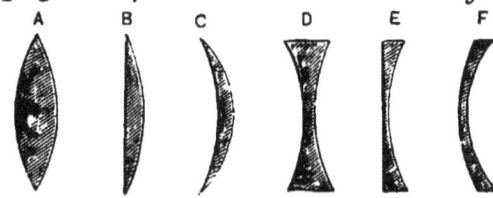

FIG. 120.—Different forms of lenses.

faces; at B one side is plane, the other convex, and the lens is called *plano-convex;* at C one side is convex, the other concave, and the lens is called *concavo-convex,* or *converging meniscus;* at D, E, and F are sections of **concave** or **diverging lenses**. D is a section of a *double-concave,* E a section of a *plano-concave,* and F a *convexo-concave,* or *diverging meniscus.*

APPARATUS.—A few concave and convex lenses; stand.

EXPERIMENT 94.—To ascertain the focus and focal length of a convex lens. In a darkened room place the lens so that a beam of parallel rays falls on it—rays from any very distant object—or it may be placed across a sunbeam. An image of the object is formed at a point F, called the *principal focus.*

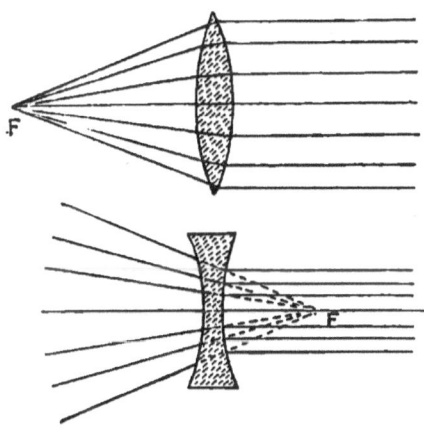

FIG. 121.—Principal focus of a double-convex, and also of a double-concave lens.

In the double-convex lens, shown in Fig. 121, the rays *actually* pass through the point F, and the focus is real.

In the case of the double-concave lens, shown in Fig. 121, the rays *appear* to come from a point F; hence the focus is *virtual.*

If u denote the distance of an object, and v the distance of the image from a mirror or lens, and if f denote the focal length, then v, u, and f are found from the formula $\frac{1}{f} = \frac{1}{v} + \frac{1}{u}$ for mirrors (as shown on p. 149), and $\frac{1}{f} = \frac{1}{v} - \frac{1}{u}$ for lenses.

It should be noted that distances measured from the lens or mirror *towards* an object are *positive*, and distances measured *away* from the object are *negative*.

In the case of the concave lens, the principal focus is on the same side of the lens as the object, and f is positive.

In the convex lens it is on the opposite side, and f is negative.

Thus distances measured from the lens or mirror in a direction *opposed* to that of the incident ray are considered **positive**, and those in the same direction as the incident ray are considered negative. The relative sizes of image and object, or $\frac{\text{image}}{\text{object}} = \frac{v}{u}$. If v and u are of the same sign, the image is real; if the signs are opposite, the image is virtual; or ratio positive image is *real*, ratio negative image is *virtual*.

If v and u are of opposite sign, the image is behind the mirror, and is therefore virtual. The ratio $\frac{v}{u}$ is called the *magnification*.

Images.—When a convex lens is used, and the object is placed between the principal focus F and the lens (Fig. 122), a virtual image A'B' is produced; the lens is then acting as a simple microscope. Thus an observer looking through the lens at an object AB, would see a virtual, enlarged, and erect image of AB. The magnifying power is the ratio of length of image to that of the object. In this manner a convex lens may be used to examine small objects of any kind, and this also explains the action of a *simple magnifying glass*.

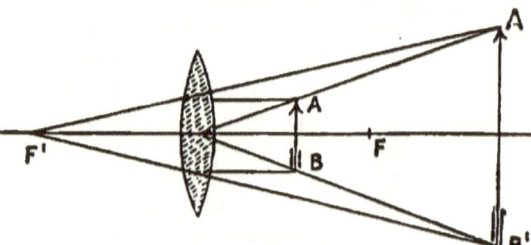

FIG. 122.—Magnifying power of a double-convex lens.

If A'B' (Fig. 123) denote an object in front of a convex lens, principal focus FO the optical centre, and FO produced the

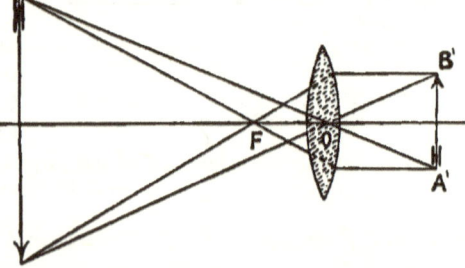

FIG. 123.—Image formed by a double-convex lens.

principal axis—if A' and B' be joined to O and produced, the image will lie somewhere between these lines. From A' and B' draw incident rays parallel to the principal axis; the refracted rays passing through F, the points A and B are determined.

The image is *real*—that is, it may be thrown upon a screen; and when a real image is formed, the image and object may be interchanged. Hence if A'B' were placed at AB, the image formed would be smaller than A'B'. The relative sizes of image and object are directly as their respective distances from the lens.

When the real image and the object are on opposite sides of the lens, the real image is always *inverted*.

APPARATUS.—A few lenses of varying focal lengths, or the large front lenses of an opera-glass or telescope, mounted in any convenient way.

Determine the focal length of a double-convex or plano-convex lens.

EXPERIMENT 95.—Focus the sun's rays on a sheet of paper; the bright spot of light is an image of the sun. Measure the distance of the lens from the image: this gives the focal length.

Obtain on a screen the image of a fish-tail burner at a distance of, say, 20 feet from a lens. It will be found that the image is real, inverted, and smaller than the flame. If the burner is brought nearer to the lens, the image increases in size; when flame and image are at equal distances from the lens, they are equal in size. When the flame reaches the principal focus, the refracted rays are parallel; and if the flame be placed between principal focus and the lens, the refracted rays would meet, if produced on the same side of the lens as the flame.

Measure the distances of the object and the image in several positions, and verify that $\dfrac{1}{v} - \dfrac{1}{u} = \dfrac{1}{f}$.

APPARATUS.—Lenses; cardboard; fine wire.

EXPERIMENT 96.—A small hole is made in a sheet of cardboard, C (Fig. 124), and across it two fine wires are attached. The hole is brightly illuminated by a gas flame; on the other side of the cardboard place the lens L and the screen S. Move either the lens or the screen until the image of the cross-wires is in focus on the screen. Measure

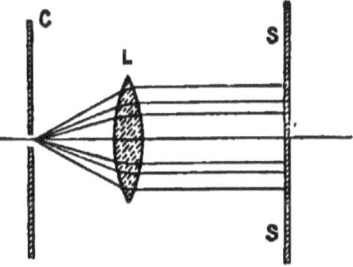

FIG. 124.—Experiment with double-convex lens.

the distance u between the cross-wires and lens, and the distance v between the lens and the screen.

Verify that $\dfrac{1}{v} - \dfrac{1}{u} = \dfrac{1}{f}$; or $f = \dfrac{uv}{u-v}$.

APPARATUS.—Concave mirror and stand; candle, or object such as an arrow.

EXPERIMENT 97.—**Concave Mirrors.** Let MNM' represent a section of a concave mirror (Fig. 125), C the centre of curvature (or centre of a sphere of which MNM' is a portion); the line NN' drawn through the centre of the mirror and through the centre of curvature is called the axis of the mirror.

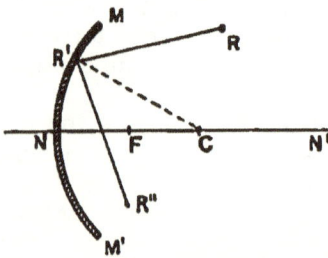

FIG. 125.—Reflection from a spherical mirror.

When a ray of light which proceeds from a point R falls on the mirror at R', the path of the reflected ray may be found by the following construction:—If C be joined to R', then R'C is the normal at R'; if the *angle of reflection* CR'R" be made equal to the *angle of incidence* RR'C, then R'R" is the path of the reflected ray. If the ray be parallel to the axis NN', then after reflection from the mirror the ray will pass through a point F, where NF = FC. F is called the **principal focus of the mirror**.

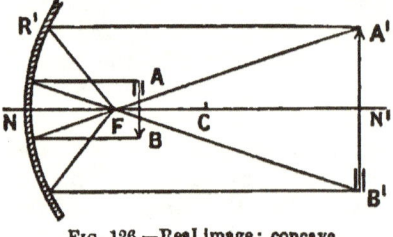

FIG. 126.—Real image; concave mirror.

Thus AB may represent a candle flame, or an object such as an arrow, which simply reflects light, may be used, and will have an image A'B', and from Fig. 126 it will be seen that the image is larger than the object, is real, inverted, and beyond the centre of curvature.

When the object is between the mirror and the principal focus (Fig. 127), the image A'B' is *erect, virtual,* and *enlarged*.

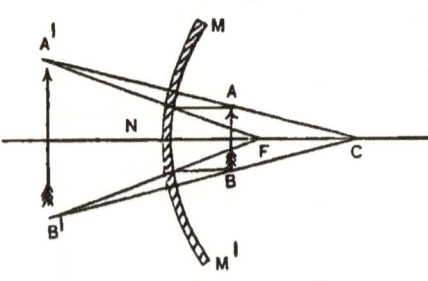

FIG. 127.

MIRRORS—LENSES.

Size of Image.—Experimentally, by placing the object at various distances, it is found that the image is smaller when placed beyond the centre of curvature, is greater when placed between the centre of curvature and the focus, and both are the same size at the centre of curvature.

All plane and all convex mirrors form virtual images; but in a concave mirror, when the distance of the object from the mirror is greater than the distance of the principal focus, the *image is real;* if the object be between the mirror and the focus, the image is virtual.

As the angle of reflection is equal to the angle of incidence, it follows that when the angle of incidence is 0°—that is, when the luminous point is placed at C (the centre of curvature)—all rays from C will be reflected back to C. Thus, if CD be a ray, the ray will be reflected back along DC; and therefore, if an object be placed at C, an image will be produced by reflection alongside it. This enables the centre of curvature of a concave lens to be found: for if the candle flame be placed on the axis of the mirror, and beyond the centre of curvature, the rays are focussed to a point on the axis between the principal focus and the centre of curvature; if the light be moved nearer to the mirror, the flame and the image approach each other until, when the flame is in such a position that the light and its image coincide, the object is at the centre of curvature. This enables the formula to be verified.

EXPERIMENT 98.—To verify the formula for a concave mirror,

$$\frac{1}{f} = \frac{1}{u} + \frac{1}{v}.$$

Make a small hole in the centre of the cardboard, and across it stretch two fine wires; place a gas or lamp flame behind it. Next, in front of the cross-wires place a concave mirror, in such a position that the image of the wires is formed by their side on the screen. The distance so determined is the radius of curvature of the mirror. If this distance be, say, 8 inches, it will be found that the distance of the image from the screen and the distance of the object from the mirror are in each case 8 inches.

Hence $\frac{1}{f}$, which is equal to $\frac{1}{6} + \frac{1}{6} = \frac{1}{3}$.

Also, to show that the focal length is half the radius of curvature, the mirror is adjusted until a well-defined image of some distant object, such as the sun, is obtained on a white screen. The dis-

150 MIRRORS—LENSES.

tance of the mirror from the screen will give the focal length; this will be found to be 4 inches.

APPARATUS.—Two glass prisms; stands.

EXPERIMENT 99.—In the previous experiments with prisms the appearance of coloured edges is sure to be noticed. The colour, which is due to the composite nature of white light, is shown as follows. Sunlight is allowed to pass through a slit S (Fig. 128) into a darkened room, and is intercepted by a glass prism G. On a screen C a long band of colour or spectrum is

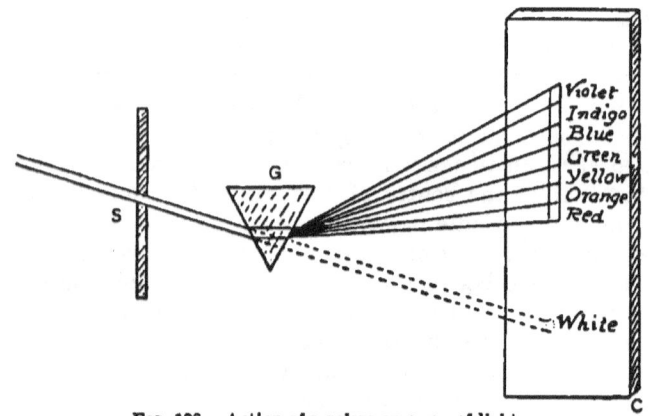

FIG. 128.—Action of a prism on a ray of light.

produced, one colour shading into the next, commencing with red and ending with violet. If a similar prism be placed so that the base of the second is nearest the edge of the first, the effect will be to produce a white spot of light, the white image of the slit, displaced owing to the refraction of the two prisms.

If a cardboard disc be divided into seven sectors and painted in the seven colours in the above order, when rotating rapidly the light from the card gives rise to a colour of white or light grey.

SUMMARY.

Plane Mirrors.—If two inclined mirrors are placed in a vertical position, the number of images of any luminous point is given by $\frac{360°}{\theta} - 1$, where θ denotes the angle between the mirrors.

Glass Wedge.—When light is refracted through a glass wedge, the rays are bent towards the base of the wedge.

Images Formed by a Lens.—Lenses may be made of glass; their curved surfaces are in many cases portions of spheres. They are usually divided into two classes—those thicker at the centre than at the edges, called *convex* or *converging lenses;* and those thicker at the edges than at the centre, called *concave* or *diverging lenses.*

Principal Focus.—When the source of light is at a *great distance*, so that *the rays are parallel*, the point to which the rays after refraction *converge* in a *convex lens*, and appear to *diverge* in a *concave lens*, is called the *principal focus.*

EXERCISES.

1. Construct the path of a ray from a luminous point to the eye in a given position, after reflection from two plane mirrors at right angles to one another.

2. In an aquarium a long tank is placed along the wall: the front of the tank is of glass; the middle of the glass is about the height of the observer's eye. Explain why the fish appear to be nearer to the observer than they really are.

3. How would you determine by experiment the path of a ray of light through a glass prism?

4. A candle is looked at obliquely in an ordinary plate-glass mirror silvered at the back. Draw the complete path of a ray from the candle, and show how to find the several images.

5. A candle is placed at a given small distance in front of a looking-glass made of thick plate glass quicksilvered at the back. A person looking obliquely into the mirror sees several images of the candle. Explain this, and show the exact positions of the images by a diagram.

6. Draw accurately the paths of four rays, two proceeding from each end of an object 2" high, placed symmetrically on the axis of a concave mirror of 4 inches focus at a distance of 6 inches from it, and thus obtain the height and position of the image.

7. Find at what distance an object must be placed in front of a concave mirror of 6 inches focal length, so that the mirror may give a real image of the object magnified six times.

$$\text{Here } \frac{1}{f} = \frac{1}{p} + \frac{1}{p^1}; \text{ but } p^1 = 6p.$$

$$\therefore \frac{1}{6} = \frac{1}{6p^1} + \frac{1}{p^1}; \quad \therefore p^1 = 7.$$

The object must be placed at 7 inches from the mirror.

8. An object 5 cm. long is placed at a distance of 40 cm. from a concave mirror of 24 cm. focal length. Find the size and position of the image.

9. A convex lens of 4½ inches focal length is held at a distance of 3 inches from a disc half an inch in diameter. Find the position and size of the image.

10. An arrow 4 inches long is placed 10 inches in front of a convex lens whose focal length is 4 inches. Illustrate by a figure, and find the position and length of the image.

11. Draw accurately the path of a ray of simple light through a 45° prism of glass, whose index of refraction is $\frac{5}{3}$, drawing the ray incident on one face in a direction perpendicular to the other face.

12. A prism of glass is placed in optical contact with a thick plate of the same kind of glass. Draw a diagram showing the course of a ray of light which passes through both the plate and the prism.

13. Two plane mirrors are inclined toward each other. Draw and explain a diagram showing how the image of an object placed between them may be seen after three reflections.

CHAPTER XIV.

HEAT—TEMPERATURE—EXPANSION OF SOLIDS.

Heat.—The word "heat" is used in physics to denote the sensation of warmth, and also that which gives the sensation. That heat is not a material substance is at once evident from the fact that a heated body weighs just the same as it did when cold. A poker placed in a bright fire becomes red-hot. Although no increase in substance has taken place, the poker is in a different condition from the former one. Its dimensions are increased, and it can communicate heat to other bodies. Heat is also produced when motion is destroyed. A few simple experiments will show this. Thus, rub a brass nail smartly on a piece of wood, hammer a piece of lead, file or saw a piece of metal—in each case heat is developed. **The effects of applying heat to a body are shown by change of size and temperature and change of state.**

Temperature.—The temperature of one body is said to be greater than that of another when, on bringing them together, the first loses and the other gains heat; and the temperatures are equal when neither gains heat from the other. Thus, temperature is a condition of bodies, and, when two bodies are placed in contact, determines which will impart heat to the other.

APPARATUS.—An iron gauge G (Fig. 129); a round zinc rod about $\frac{1}{4}$ inch in diameter, and made in the form of a T-piece, with a wood handle at H.

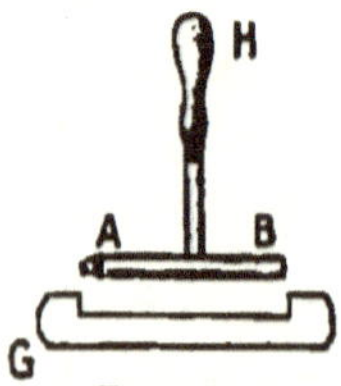

EXPERIMENT 100.—A zinc rod A B is made to fit tightly between the two ends of an iron gauge G (Fig. 129). If this gauge be heated by means of a Bunsen burner, the rod will be found to fit quite loosely; but if the rod be heated, it increases in length, and cannot be put within the gauge. Again, if both be heated by plunging into boiling water, it will be found impossible to place the rod in position.

FIG. 129.
Linear expansion.

HEAT—TEMPERATURE—EXPANSION OF SOLIDS. 153

It is found by experiment that a bar of any material will expand *a certain small fraction of its length (called the coefficient of linear expansion) for every degree Centigrade which it is heated.* For zinc this fraction is ·000029, for iron ·000012.

Thus the expansion of the zinc rod for 1° C. = ·000029, for 80° C. = 80 × ·000029 = ·00232; and if the original length were 10 inches, the increase in length = 10 × ·00232 = ·0232 inches, and actual length of rod = 10·0232 inches.

In a similar manner the alteration in length of the iron gauge can be estimated. The fraction for brass is ·000019, or $\frac{19}{12}$ that of iron. That the expansion of brass is greater than that of iron may be shown by the following experiment.

APPARATUS.—Two strips iron and brass respectively, each about 2 feet long, riveted together.

EXPERIMENT 101.—Hammer the compound bar straight, as shown by A (Fig. 130). Heat the bar by a flame such as a Bunsen. When heated the bar will be found to be bent as shown at B, the brass expanding more than the iron. The bar is curved as shown, the strip of brass being on the outer or convex part.

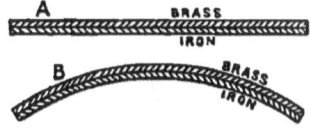

FIG. 130.—Experiment to show unequal expansion.

Familiar instances of expansion are numerous—such as the cracking of a lamp glass, or of a glass tumbler into which boiling water is poured. In both cases the heat applied to one part of the glass does not pass readily to the other parts, on account of glass being a bad conductor of heat; the heated portion expands, and rupture occurs. To avoid breakage due to unequal expansion, thin flasks and sand-baths are used in a laboratory.

To allow for expansion of iron or steel rails by heat, spaces are left between the ends of consecutive rails. In the Forth Bridge (1,700 feet span) a total space of about 18 inches is allowed.

Standard yard measures and **metre scales** are constructed so that they are the required length at some given definite temperature.

The English standard yard is the correct length when the temperature is 62° F. or $16\frac{2}{3}$ C., and the metre when the temperature is at 0° C. If used at a temperature differing to any appreciable extent from that at which they were graduated, the length will not be the same. Thus for accurate estimations corrections have to be applied to all brass or glass scales when these are attached to barometers, etc. It is important to be able to estimate the alteration in length due to any given rise of temperature.

Thus if l denote the original length of a bar, θ the number of degrees through which its temperature is raised, and c the coefficient of expansion, when the bar has been raised through a number of degrees denoted by θ, then length $= l(1 + c\theta)$.

Values of c for a few well-known materials are as follows:—

TABLE IV.

Material.	c or coefficient of linear expansion per 1° C.	Coefficient of cubical expansion per 1° C.
Iron	·000012	·000036
Copper	·000017	·000051
Zinc	·000029	·000087
Brass	·000019	·000057
Glass	·0000089	·0000267
Platinum	·0000089	·0000267

It will be noticed that the value of c is usually different for various materials. In the case of platinum and glass the values are alike; hence the following experiment can be tried:—

APPARATUS.—Glass tubing; platinum wire.

EXPERIMENT 102.—Fuse a piece of platinum through the side of a glass tube; on cooling it will be noticed that the glass does not crack.

APPARATUS.—Flat rod of metal about 12 inches long; wooden blocks; Bunsen burner.

EXPERIMENT 103.—Lay the bar across the two wood blocks as shown in Fig. 131. Fasten one end of the bar by placing a weight on it, and under the other end of the bar, at right angles to the length of the rod, place a fine needle; through the eye of the needle pass a fine straw (or strip of aluminium foil), to act as a pointer

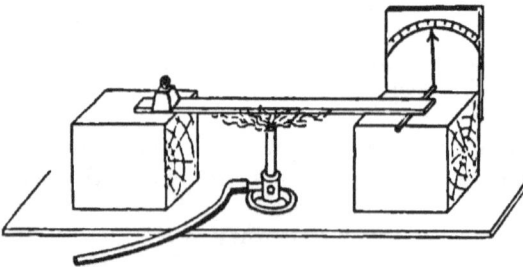

FIG. 131.—Linear expansion of a metal bar.

—this should be fastened at right angles to the needle by means

HEAT—TEMPERATURE—EXPANSION OF SOLIDS. 155

of wax. When the bar is heated and expands, it rolls the needle. A slight motion of the needle is shown by a much larger distance moved through by the end of the pointer. By means of a cardboard scale as shown the alteration in length can be determined. When the bar cools, the pointer moves back to the starting-point. If c denote the coefficient of expansion, to find the expansion of a square of unit side (1 cm. or 1 inch), $(1 + c)^2 = 1 + 2c + c^2$. Now as c is small, c^2 is very small compared with $2c$, and may be neglected; hence the coefficient of expansion for a square is $2c$, or twice the linear.

The small fraction which a unit volume of any material is found to expand when raised from 0° C. to 1° C. is called *the coefficient of cubical expansion,* and *is* (practically) *three times the coefficient of linear expansion.*

Thus $(1 + c)^3 = 1 + 3c + 3c^2 + c^3$.

As c is very small, the term c^2 is very small, and c^3 is smaller still; hence the two last terms may be neglected, and the coefficient of cubical expansion is $3c$, or three times the linear.

APPARATUS.—Gravesande's ball and ring; Bunsen burner.

EXPERIMENT 104.—To show the expansion of solids when heat is applied, we may take a brass ball (Fig. 132), and a ring through which the ball will pass quite easily when ring and ball are at the same temperature. If the ball be heated by the flame of a Bunsen burner or by plunging it into boiling water, it will not pass through the ring, but will rest upon it—showing the expansion due to heat. If allowed to rest upon the ring for a short time, the ball cools and contracts, while the ring becomes heated, and therefore enlarges, and the ball readily passes through.

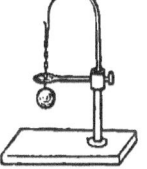

FIG. 132.
Expansion of a metal ball.

APPARATUS.—Sheet of tin; glass flask.

EXPERIMENT 105.—Out of a piece of tinplate cut a hole just large enough to allow a round-bottomed thin glass flask to pass through. If now the flask be filled with boiling water, it will not pass through; but if held in contact with the plate, it will pass through when the two are at or near the same temperature.

SUMMARY.

Heat is not a material substance.
Temperature of a body indicates its condition or thermal state considered with reference to its power to give out or to receive heat from other bodies.
Effects of heat are (1) to cause expansion, the amount varying with

156   HEAT—TEMPERATURE—EXPANSION OF SOLIDS.

different materials (when allowed to cool, the original dimensions are obtained); (2) to raise the temperature; or (3) to change the state, such as to convert a solid into liquid or liquid into a gas—that is, ice, water, steam.

Glass and platinum expand equally.

The *superficial expansion* of a substance is twice its linear, and the *cubical expansion* is three times its linear expansion.

EXERCISES.

1. What is meant by the statement, The coefficient of linear expansion of copper is ·000017? If a copper rod is 20 yards long at 0° C., how much longer will it be at 90° C.?

2. If an iron yard measure be correct at the temperature of melting ice, what will be its error at the temperature of boiling water?

3. The coefficient of linear expansion of brass is 0·000019; what will be the volume at 100° C. of a mass of brass the volume of which is 1 cubic decimetre at 0° C.?

4. If the mean coefficient of expansion of water between 0° C. and 50° C. is ·000236, find the weight of water displaced by a brass cube whose side is 1 cm. in length at 0° C. when the cube and the water are both at 50° C.

5. Find the difference in the length of an iron railroad between two places 400 miles distant (London and Edinburgh), in winter and summer, the temperature difference being 21° C.

6. A light lath a metre long can turn about a hinge at one end. A piece of wire 2 metres long is attached to the lath at a distance of 1 cm. from the hinge. The wire is vertical, and its upper end is fixed in such a way that the lath is horizontal. On heating the wire 50° C. the end of the lath sinks through 16 cm.; find the coefficient of expansion of the wire.

CHAPTER XV.

EXPANSION OF LIQUIDS AND GASES.

Expansion of Liquids.

APPARATUS.—Glass flask, tubing, cardboard, and water. Fit the flask with a tight-fitting cork and long glass tube passing through it: a scale may be attached by making two slits in the scale, threading the tube through them. The tube must pass through the cork and project a short distance, so that the bottom of the tube is below the surface of the liquid in the flask.

EXPERIMENT 106.—Before putting the tube in position, fill the flask with water to which a little colouring matter has been added. Press the cork into the flask until it is fairly tight; the water will rise in the tube—note its height. Next plunge the flask into a vessel containing water at a temperature of 30° C. The expansion of the glass will cause a momentary sinking of the water in the tube, after which the water will rise. When the expansion ceases, and the column no longer ascends, note the reading.

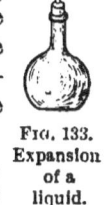

FIG. 133. Expansion of a liquid.

Repeat the experiment when the temperatures are 40°, 50°, etc., in each case noting the heights to which the liquid ascends. Tabulate as follows :—

Temperature.	Reading.

EXPERIMENT 107.—Fit three corks into the necks of three small flasks of equal size, and a narrow glass tube open at both ends tightly into each cork. Fill the flasks with mercury, coloured water, and methylated spirit respectively. Push in the corks so that the liquid stands at the same height in the three tubes. Fix a paper or cardboard scale behind each, and place all three flasks in a convenient vessel. Pour into this sufficient warm water to cover the flasks up to the corks. The expansion of the glass will cause a momentary sinking of the columns, after which expansion takes place. When the expansion ceases, and the flasks are all at the same temperature, it will be found that the mercury has risen least, the water next, and the methylated spirit most of all.

Absolute and Apparent Expansion of a Liquid.—In order to apply heat to a liquid, it is necessary to place it in a vessel of some kind of solid substance; and when heat is applied, both vessel and liquid expand. The increase in volume of the liquid (due to expansion of the vessel and of the enclosed liquid) is observed, and is called the *apparent expansion* of the liquid, the apparent amount being evidently less than the actual amount. If the expansion of the vessel be allowed for, we obtain the real expansion of the liquid.

Expansion of Gases.

APPARATUS.—Small flask fitted upside down, as shown in Fig. 134; a piece of glass tubing passing through a well-fitting cork into the flask, and the lower end of the tube just below the surface of the water in the beaker, as shown—the water in the beaker A has a little colouring matter added to it.

EXPERIMENT 108.—Heat the flask B; which may be effected by pouring hot water over it. When this is done, bubbles of air will be seen in the vessel A. Allow the flask to cool, and the water will rise in the tube; the height of the column can be noted by the scale attached. The instrument forms a fairly good *thermoscope*. If water colder than the air in the room be poured on the flask, the column will ascend; if warmer, the column will sink. Thus, by means of the scale attached to the tube, and noting the position of the liquid column when the air was heated by different bodies, the instrument may be used to compare temperatures; hence it is called an **air thermometer.**

FIG. 134.—The principle of the air thermometer.

Instead of the water column, a small column of mercury may

EXPANSION OF LIQUIDS AND GASES. 159

be used. If the flask B be slightly heated, and the open end of the tube placed in clean mercury, a small column can be enclosed, the position of the column indicating the pressure of the air in the flask B. For any specified temperature difference, a volume of air expands about 20 times as much as an equal volume of mercury; hence the distance between what are called the *fixed points in an air thermometer* would be about 20 times the distance in a mercurial thermometer.

Equal Expansion of Gases.—In the preceding experiments we have found that both solids and liquids expand by heat, and also that the expansion for any temperature difference is not the same for either different solids or different liquids.

In the case of gases, for any alteration in temperature the corresponding alteration in volume is practically the same for all gases, and is $\frac{1}{273}$ of their volume at 0° C. for each increase of 1° C. Thus a cubic foot of gas at 0° C. would become 2 cubic feet at 273° C.

The relations between volume, pressure, and temperature are expressed in laws, and the two important laws may be referred to here :—

Boyle's Law.—*If the temperature of a gas be kept constant, the volume is inversely proportional to the pressure;* or, $p \times v =$ constant.

Charles's or Dalton's Law.—*If the pressure be kept constant, the volume of a given mass of gas increases by a definite amount for each degree of rise in temperature.*

If v_t and v_o represent the volume of the gas at $t°$ and $0°$ respectively, then the law may be written as $v_t = v_o \left(1 + \dfrac{t}{273}\right)$.

If $p_1 v_1 t_1$ represent the pressure and volume of a given mass of gas at a temperature t_1, and $p_2 v_2$ the pressure and volume at another temperature t_2, the *two laws* are included in

$$\frac{p_1 v_1}{1 + \dfrac{t_1}{273}} = \frac{p_2 v_2}{1 + \dfrac{t_2}{273}}$$

APPARATUS.—Two small flasks, corks, and tubing.

EXPERIMENT 109.—Two small flasks equal in size are fitted with corks, and with narrow tubes bent so that they will dip under water in a shallow vessel. Fill one flask with air, the other with coal-gas; also fill two equal test tubes with water, and invert them over the ends of each tube.

Pour warm water into the vessel until the flasks are covered up to the corks. Bubbles of gas will rise up the test tube, and will displace the water. The increase of volume of the coal-gas in one flask, and of the air in the other, by the application of heat, is indicated by the amounts of gas collected in the two tubes; these amounts will be found to be equal. If the experiment be repeated, using different gases, such as oxygen, hydrogen, etc., the same result will be obtained.

Differential Thermometer.

APPARATUS.—A U-shaped piece of glass tubing connected, as shown in Fig. 135, to two bent tubes terminating in globes or flasks. The apparatus is easily made by bending a straight tube as shown for the centre, and using two pieces of india-rubber tubing to connect to the bent tubes attached to the globes A and B.

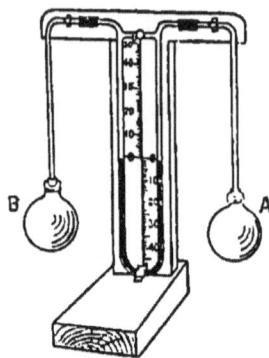

FIG. 135.—Differential air thermometer.

EXPERIMENT 110.—The U-tube in the centre contains coloured water, filling each limb to a height of about 2 inches. It may be used to ascertain the difference between the temperature of a liquid and that of the air in the room. Thus, if the liquid is made to cover the flask A, then if it be warmer than the air, the air contained by A will expand, and the column will descend on the side nearer A and rise on the opposite side; the converse will take place if the liquid is colder. If the temperature is the same, the column remains at rest. In a similar manner, the apparatus furnishes an easy method of ascertaining any slight difference in the temperature of two liquids.

SUMMARY.

Liquids.—As in the case of solids, the effect of heat is to cause expansion, the expansion varying with different liquids.

Gases.—The effect of heat is to increase the volume for the same temperature difference. The expansion of a given volume of gas is more rapid than that of an equal volume either of a solid or a liquid.

Equal Expansion of Gases.—Gases which are far removed from their point of liquefaction expand $\frac{1}{273}$ of their volume at 0° C. for every rise of 1° C.

Boyle's Law.—If the *temperature* be constant, the *pressure* of a given quantity of gas is *inversely proportional to the volume*.

Charles's or Dalton's Law.—If the *pressure* be constant, a gas *expands by* ·00366 *of its volume* at 0° C. when its *temperature is raised* 1° C.

EXPANSION OF LIQUIDS AND GASES.

EXERCISES.

1. Distinguish between the absolute and the apparent expansion of a liquid, and show that the coefficient of absolute expansion is equal to the sum of the coefficient of apparent expansion, and of the coefficient of expansion of the vessel.

2. The apparent coefficient of expansion of mercury in a glass vessel = ·000153, from Table IV.; the cubical expansion of glass = ·0000267; ∴ absolute coefficient of expansion of mercury = ·000153 + ·0000267 = ·00018.

3. A cubical vessel whose edge is 1 foot is made air-tight when the barometer stands at 30 inches and the temperature of the air is 15° C.; if the temperature of the air is raised to 60° C., what is the increase of the pressure of the contained air on each face of the cube?

(*N.B.*—The coefficient of expansion is 0·003665, and a cubic inch of mercury may be taken to weigh half a pound.)

4. A certain quantity of gas is contained in a vessel whose volume is one cubic foot and its temperature is 20° C. If in any way (for example, by pressing down a piston) its volume is changed to 1,000 cubic inches, and its temperature raised to 30° C., find the ratio of the pressure of the gas in its former state to its pressure in its latter state.

(*N.B.*—The coefficient of expansion is 0·003665.)

5. The apparent expansion of a liquid when measured in a glass vessel is ·001029, and when measured in a vessel of copper is ·001003. If the coefficient of linear expansion of copper be ·0000166, find that of glass.

CHAPTER XVI.

THERMOMETERS.

Thermometers.—The usual method of comparing **temperatures** is by means of **thermometers** and **pyrometers.** This enables the degree of heat to which a body has been raised to be observed, but *does not* indicate the *quantity of heat in the body.* When a small quantity of water is placed in a vessel, and heat applied by means of a Bunsen burner, the rise of temperature can be noted at the end of equal intervals of time. If a much larger quantity of water be subjected to the same amount of heat, and the temperature at the same intervals as before be noted, they will be found not to be equal to those obtained in the first instance, although, roughly, the same amount of heat has been given.

If the contents of two vessels, one containing water at a lower temperature than the other, be mixed, *no heat is lost by the mixing of the two;* but the thermometer will show a gain as compared with the previously observed temperature of the colder, and a loss as compared with the hotter. Hence the thermometer *does not* indicate the quantity of heat.

Nearly all bodies expand by heat and contract by cold, and the usual method of measuring temperature is by the expansion or contraction of some liquid.

In Experiment 107 it has been found that liquids when heated expand; by noting the divisions on the scale to which the liquid rises, a rough kind of thermometer is obtained.

To Make an Alcohol Thermometer.

APPARATUS.—Thermometer tubing; Bunsen burner; alcohol coloured with aniline magenta.

EXPERIMENT 111.—Soften one end of a piece of tubing in the Bunsen flame, blow a strong bulb about $\frac{3}{4}$ inch in diameter on one end, and allow the tube to cool. To introduce the alcohol, heat the bulb carefully over a flame and immerse the open end in the coloured alcohol—a small quantity will be introduced; connect the end of the tube by a short piece of india-rubber tubing to a glass funnel, and place some alcohol in the funnel; warm the bulb, thus expelling some of the air, which escapes in bubbles through the alcohol; remove the flame and cool the bulb, and the alcohol enters the tube.

Boil the alcohol in the bulb until nearly all is boiled away. The air is driven out by the vapour of the alcohol. Then allow to cool, and the alcohol enters the bulb and fills it. The range of the instrument should be about a distance of 1 inch for a rise in temperature of 10°. Ascertain by immersion in hot water if this is the case; if not, another bulb must be prepared, larger or smaller as may be required, to give about this range. Prepare an index of a short thread of black glass of a size which moves easily in the tube. Immerse the thermometer in a bath of water a little warmer than the highest temperature to be registered. If the tube is not quite full of liquid, more must be added to fill it completely. Seal the end with a small blow-pipe flame, and after removing from the bath complete the sealing, making the end as strong as possible. Mark the freezing-point as shown in Fig. 137. Immerse the alcohol thermometer, together with a mercurial thermometer, in a bath of water sufficiently warm to bring the alcohol nearly to the end of the tube. Make a series of equidistant marks on the tube, the distance between them about half an inch. As the water cools, read the temperatures at the instant the liquid passes each mark. On a sheet of squared paper plot a series of points to represent the temperatures corresponding to the marks. Draw a curve evenly through the points to correct errors of observation, and transfer the scale to the piece of paper pasted on the board to which the thermometer is to be attached. It will be found that the distances corresponding to one degree are greater the higher the temperature.

In thermometers the liquid commonly used is *mercury;* its special advantages for this purpose are as follows:—

1. It remains liquid through a long range of temperature. Its freezing-point is about − 40° C., and its boiling-point 350° C.

2. Its constant rate of expansion.

3. Its low specific heat. It therefore takes little heat from the body tested; which allows very slight changes in temperature to be registered.

THERMOMETERS.

4. It is a good conductor of heat, thus readily acquiring the temperature of a body.

5. It is opaque, so that the motion of the column can easily be followed.

6. It does not wet the glass or cling to it, but moves easily and rapidly along the tube; by this means changes in the height can be quickly and accurately read off.

When the temperatures are low, alcohol is used instead of mercury. Alcohol does not freeze until the temperature is reduced to about $-130°$ C., and it is in consequence suitable for such purposes. For very high temperatures metal (platinum) is used.

Fixed Points of a Thermometer.—Two thermometer tubes are shown in Fig. 136, formed by blowing a bulb at the end of a capillary tube, or a tube with a fine bore. The mercury cannot be poured into the bulb, but by gently heating some of the air is expelled; and if the open end be placed under mercury, a little will be drawn up the tube as the tube cools: in this manner, by alternately heating and cooling, the tube and bulb can be filled with mercury. When full, the end of the tube is sealed at a temperature higher than the thermometer is intended to register; the instrument is now put aside for some time, before what are known as the *fixed points* are marked.

The bulb when formed does not contract at once to the final dimension, but shrinkage goes on gradually for some time after being made; hence good thermometer tubes when made are laid aside for a considerable period before being filled.

Fig. 136. Thermometers.

What are called the fixed points of a thermometer are the **freezing** and **boiling** points of water. Thus the temperature at which ice melts or water freezes is found to be always the same at the same pressures, and at normal pressure the temperature of steam from boiling water is also constant.

Freezing-Point.—The thermometer is inserted into a vessel containing lumps of (pure) melting ice (Fig. 137), and allowed to remain until the column of mercury eventually comes to rest; the exact height of the column is then marked by a scratch on the tube at the point where the mercury stands.

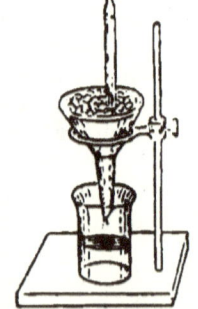

Fig. 137. Marking the freezing-point.

Boiling-Point.—This is obtained by placing the thermometer in *the steam* given off by water boiling *under the standard atmospheric pressure*, the height of the mercury column being marked by a scratch on the glass as before.

Graduating a Thermometer.—For comparison of observations and experiments some definite value must be attached to the two fixed points, and the stem of the instrument must be graduated. These graduations are called *degrees*.

In **the Fahrenheit thermometer** the *freezing-point* is marked 32°. The interval between the freezing and boiling points is divided into 180 equal parts or degrees.

In **the Centigrade thermometer** the freezing-point is marked 0°, and the interval between the fixed points is divided into 100 equal parts; hence boiling-point is marked 100°.

In **the Réaumur** the freezing-point is marked 0° and the boiling-point 80°.

The Fahrenheit and Centigrade are the two usually used, and readings are indicated by the addition of F. or C. to the reading—thus, 32° F. or 0° C.

The three thermometers are shown in Fig. 138.

The Centigrade is usually adopted for scientific purposes.

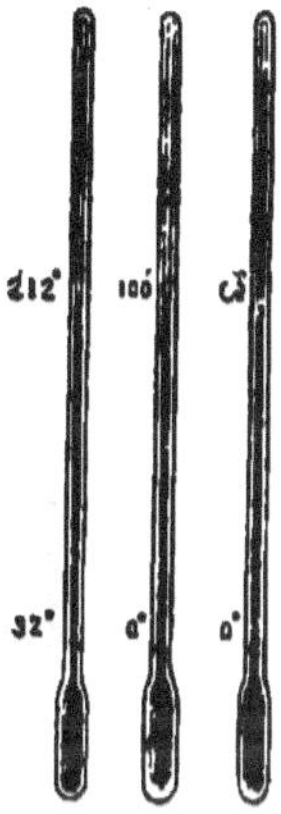

To indicate temperatures below freezing-point on the Centigrade thermometer, a minus sign is prefixed. Thus − 8° C. would indicate 8 degrees below freezing-point on the Centigrade scale.

FIG. 138. Fahrenheit, Centigrade, and Réaumur thermometers.

If the number of degrees indicating the two fixed points in the three systems be carefully remembered, it is easy to convert any temperature reading from one scale to any other.

As will be seen from Fig. 138, the reading 100° C. = 180° F. + 32; thus,—
1° C. = $\frac{5}{9}$(F° − 32).

Thus, 98° F. = $\frac{5}{9}$(98 − 32) = 36$\frac{2}{3}$° C.

∴ To convert from the Fahrenheit to the Centigrade scale, subtract 32 from the reading and multiply by $\frac{5}{9}$.

Also 50° C. = $\left(50 \times \frac{9}{5}\right)$ + 32 = 122° F.

∴ To convert from the Centigrade to the Fahrenheit scale, multiply by $\frac{9}{5}$ and add 32.

It must be noticed that the so-called fixed points are fixed only in the sense that the conditions under which they were obtained are always complied with. Thus a reduction of the standard atmospheric pressure would lower the temperature at which boiling occurs; for example, it is found that the boiling-point is lowered 1° C. for every 1,000 feet of ascent. Therefore at a height of 5,000 feet above the sea-level the boiling-point would be 95° C. approximately.

APPARATUS.—Round-bottomed 4-oz. flask; Bunsen burner; retort stand; beaker.

EXPERIMENT 112.—Boil water in the flask, and, when all the air is expelled, cork. Invert the flask (Fig. 139), and allow it to cool for a short time. The water vapour which mainly filled the space above the water condenses, and the pressure upon the water diminishes. If now water be poured on the top, as shown, the water will boil under the reduced pressure.

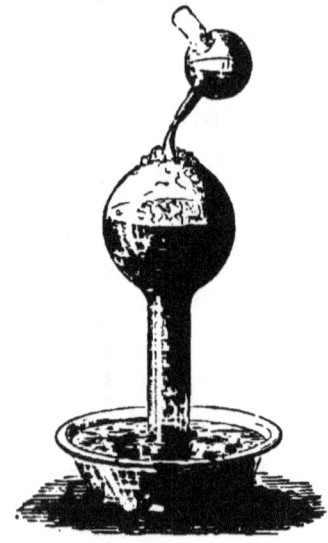

FIG. 139.—Boiling-point under reduced pressure.

A similar result can be observed if the flask which contains the water be placed uncorked under the receiver of an air-pump. If the air be quickly withdrawn, and therefore the pressure reduced, bubbles of steam begin to form, and ebullition occurs.

When water containing any chemical compound is lowered sufficiently in temperature so that ice is formed, a thermometer placed in the water will not indicate 0° C., but a lower temperature; hence the importance of using clean ice in fixing the freezing-point of a thermometer.

EXPERIMENT 113.—Mix some common salt with water and ice; insert a thermometer, and note that the temperature falls below 0° C.

Hope's Experiment.—In the preceding experiments it has been found that both solids and liquids expand when heated, but water at the temperature 0° C. is found to contract when heated, until the temperature reaches 4° C., when it begins to expand. Hence water is said to have its maximum density at 4° C.

Hope's apparatus consists of a metal cylinder, C, with a trough-shaped cavity at the centre in which a freezing mixture can be placed (Fig. 140). Thermometers are inserted, as shown at A and B. By these the temperature of the water at the top and bottom of the cylinder can be ascertained.

EXPERIMENT 114.—The vessel is filled with water cooled down to about 8° C., and a freezing mixture placed in the trough T; the effect of the freezing mixture is to cause the lower thermometer to fall until it registers 4° C., the upper one being scarcely affected. When the lower thermometer registers 4° C., the upper one begins to fall, and continues to do so until a temperature 0° C. is reached. The lower thermometer does not fall to 0°, but stops at 4°; thus the water at 0° is lighter than the water at 4°, and floats upon its surface.

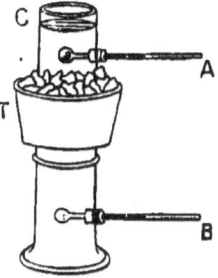

FIG. 140.
Hope's experiment.

Enter all the particulars of the experiment in your note-book, and also any conclusions you are able to draw from it.

SUMMARY.

Thermometers and Pyrometers are used in comparing and measuring temperatures.

Mercury is chiefly used in thermometers for all ordinary temperatures.

Advantages due to using Mercury.—Remains liquid through a long range of temperature. Its low specific heat. A good conductor of heat. Does not wet or cling to the glass. It is opaque, and can readily be seen.

Fixed Points.—The *boiling* and *freezing* points of water under normal pressure are taken as fixed points. The distance between the fixed points is divided into a scale or number of equal parts or divisions.

Centigrade Scale.—Freezing-point, 0° C.; boiling-point, 100° C.

Fahrenheit Scale.—Freezing-point, 32° F.; boiling-point, 212° F.

Maximum Density of Water is at a temperature of 4° C.

Change of Pressure.—The boiling-point is the temperature of steam under a normal pressure; it is *lowered* by *diminishing the pressure*, and *raised* when the *pressure is increased*.

EXERCISES.

1. Find the Centigrade temperatures corresponding to the following Fahrenheit temperatures—374°, 365°, 320°, 71·60°, 23°, 14°.

2. Find the Fahrenheit temperatures corresponding to the following Centigrade temperatures—260°, 242°, 200°, 169°, 121°, −1°, −7°.

3. A thermometer is graduated so that it reads 15° in melting ice and 60° in normal steam. Convert into Centigrade degrees the readings 20° and 90° taken on the thermometer.

4. A temperature is 20° on the Fahrenheit scale; what is it on the Centigrade scale?

5. How is a mercurial thermometer graduated?
If you add 32 to the sum of the readings on the Centigrade and Réaumur thermometers, show that you invariably arrive at the reading on the Fahrenheit thermometer.

6. Explain exactly the nature of boiling. Is it possible to make lukewarm water boil without heating it; and if so, how?

CHAPTER XVII.

SPECIFIC HEAT—CHANGE OF STATE.

Specific Heat.—That the amount of heat required when equal weights or equal volumes of different substances are heated through equal intervals of temperature is different, can be shown by the following experiment.

APPARATUS.—A few metal balls, of iron, lead, and copper; a cake of beeswax about $\frac{1}{4}$ inch in thickness, supported either by a retort stand or by a wire frame, as shown in Fig. 133.

EXPERIMENT 115.—Heat the balls to about 150° C. in the oil-bath, and drop them simultaneously on to the wax by means of a small wire frame, W (Fig. 141). It will be found that the iron quickly melts its way through the wax, the copper takes a longer time, while the lead only penetrates a short distance into it. The lead ball is heavier than the iron, but does not get through as quickly as the iron; hence the iron gives out more heat in cooling than the lead; or, the capacity for heat is greatest in the iron, is less in the copper, and is least in the lead. If tin, bismuth, etc., had been used, similar results would have been obtained. Thus every substance has a distinct *specific heat*, which is measured by *the quantity or number of units of heat required to raise the temperature of unit mass of the substance through 1° C.*

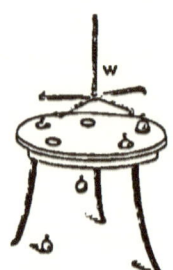

FIG. 141. Experiment to show capacities of different solids for heat.

(It should be noted that the result obtained is only a rough approximation. The conductivity of the metals affects the rates at which the balls pass through the wax, also the masses are different, and therefore the metals do not always drop through in the order of their specific heats.)

SPECIFIC HEAT—CHANGE OF STATE.

Capacity for Heat or Thermal Capacity of a Substance.—Every substance has a distinct specific heat S, and the product of the specific heat and the mass is called the thermal capacity. Hence, if the specific heat of a body of mass M be denoted by S, the *capacity for heat is MS*.

Specific Heat of Lead.

APPARATUS.—Into the cork fitting the neck of a flask a test tube, t, is inserted (Fig. 142); also a piece of bent tube S, through which steam can escape. Bunsen burner, retort stand, lead shot, beaker, and thermometer.

EXPERIMENT 116.—Weigh into a beaker a quantity of water having a temperature T_2, and a weight W_2. Also place a known weight of lead shot, W_0, into the test tube t. Heat the lead to the temperature of the steam (100° C.); this is easily done by boiling the water for some time (about a quarter of an hour). Put the hot lead into the beaker, and gently stirring the water with the thermometer, ascertain its temperature, T_3. The lead has cooled from 100° to T_3, and the water has increased in temperature from T_2 to T_3. Hence if S denote the specific heat of lead, then

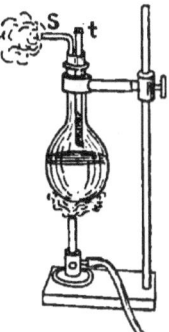

FIG. 142.—Specific heat of lead.

$$S = \frac{W_2(T_3 - T_2)}{W_0(100° - T_3)}.$$

To prevent loss of heat by conduction and radiation, the beaker may be put into a larger one, and the space between the two lined with cotton wool.

EXAMPLE.—100 grams of lead shot are heated to 100° C., and the lead is then put into a beaker containing 100 grams of water at a temperature 14° C.: the resulting temperature is found to be 16·5° C.; find the specific heat of the lead.

The lead has cooled from 100° to 16·5°; ∴ (100° - 16·5°) = 83·5°.

If S denote the specific heat of the lead, then heat given out by lead = 100 × S × 83·5. But the heat given out by the lead has raised the temperature of 100 grams of water from 14° to 16·5°, or 2·5°;

∴ S × 100 × 83·5 = 100 × 2·5.

$$S = \frac{250}{8350} = ·029.$$

Repeat the experiment, using short pieces of copper and iron wire, also mercury, etc. In each case the heat abstracted by the beaker, thermometer, etc., has been neglected, and the results obtained are only approximate. The heat abstracted by the beaker or calorimeter in this and the following experiments may be estimated as shown in the example on page 174.

Enter carefully the results in your note-book, and afterwards compare with Table V.

TABLE V.

Material.	s	Material.	s
Lead	0·03	Mercury	·033
Copper	·095	Tin	·056
Iron	·114	Zinc	·095
Glass	·2	Ice	·5

Specific Heat of Glass.

APPARATUS.—Beaker; a quantity of finely-broken glass.

EXPERIMENT 117.—Weigh a beaker nearly filled with finely-broken glass. Prepare some hot water at a known temperature, T_0. Pour the hot water into the beaker and note the temperature, stirring gently with the thermometer; let the resulting temperature be T_3. Tabulate as follows:—

W_2 = Weight of glass at temperature T_2.
W_0 = Weight of water at temperature T_0.
Resulting temperature = T_3.

Then heat gained by glass = heat lost by water;
or, weight of glass × specific heat × rise in temperature
= weight of water added × fall in temperature.

∴ specific heat × $W_2(T_3 - T_2) = W_0(T_0 - T_3)$.

$$\therefore \text{specific heat} = \frac{W_0(T_0 - T_3)}{W_2 T_3 - T_2}.$$

The specific heat is found to be about ·2.

Change of State—Fusion.—The preceding experiments have shown that in general the application of heat to a solid causes a rise in its temperature and an increase in size. In the case of nearly all solids, if the temperature be raised sufficiently, they are found to change from a solid to a liquid state. This change is called *fusion*, and the temperature at which it occurs is called the *melting-point*. When the solid begins to melt, the temperature does not alter until *fusion* is complete; but when completely melted, the temperature again rises on the further application of heat.

Change of State—Melting-Point.

APPARATUS.—Small flask ; ice ; tubing.

EXPERIMENT 118.—Fill a flask with pounded ice and water (Fig. 143). Through a well-fitting cork a straight tube is inserted, and the cork pushed in until the liquid stands a certain height in the tube. The height can be read off by the scale attached. Apply heat to the flask (which may be done by putting the flask in a vessel of lukewarm water, or by using a Bunsen burner turned low). As the ice melts, the height of the liquid in the tube will become less. This contraction will be followed by expansion after a temperature of 4° has been reached, if the supply of heat be maintained.

FIG. 143. Expansion of water on freezing.

Melting-Point of Beeswax.

APPARATUS.—Glass tubing ; beeswax.

EXPERIMENT 119.—Draw out a piece of glass tubing or a test tube into a fine tube by the blowpipe. Cut off a piece 2 or 3 inches long, and seal up one end in the flame. A few pieces of wax are inserted, and the tube tied to a thermometer in such a manner that the wax is near the bulb of the thermometer. The two are now inserted in a beaker of water and heat is applied gradually; then the *temperature* at which the first drop runs down the tube can be observed ; or the tube previously opaque with the solid wax becomes transparent when the wax melts, and the temperature at which this occurs is noted. The water is now allowed to cool, and the temperature at which the wax solidifies and the tube again becomes opaque is observed. The mean of the two readings will give the *melting-point* required.

Melting-Point of Paraffin.

EXPERIMENT 120.—Fill a short length of glass tube of small bore with melted paraffin, and when cold cut off a piece about an inch long and tie it to the bulb of a sensitive thermometer. Immerse the tube and thermometer in a beaker of cold water. By means of a small lamp or Bunsen flame gradually warm the water, and note the temperature when the paraffin is just melted. Allow the water to cool, and observe the temperature at which the paraffin solidifies. The mean of the two readings will be the melting-point of paraffin (about 50° C.).

Latent Heat.

APPARATUS.—Beaker, retort stand, Bunsen burner, ice, and a thermometer.

EXPERIMENT 121.—Put some pounded clean ice in a beaker and insert a thermometer. Place the beaker on a retort stand, and apply heat gradually by means of a Bunsen burner turned low. As the ice gradually melts, watch the thermometer. It will be found that the temperature does not rise, but remains at 0° C. so long as any ice is present. When all the ice is converted into water, the thermometer will indicate a rise in temperature, and will continue to do so until the temperature reaches 100° C., when the water will boil and steam will be formed. The thermometer will not indicate any further rise, although the supply of heat is maintained.

The heat applied to change the ice into water and the water into steam, and which is not indicated by the thermometer, is called *latent heat*. Thus *the latent heat of water* is the number of units of heat required to convert 1 gram (or 1 lb.) of ice at 0° C. into water at the same temperature.

APPARATUS.—Beaker; ice; Bunsen burner.

EXPERIMENT 122.—The beaker may, in order to prevent excessive loss by radiation and conduction, be enclosed in a larger one, and the space between the two lined with cotton wool. Weigh the beaker or beakers, and run in water of weight W_2 and temperature T_2.

Weigh out n grams of ice. This should be dried as quickly as possible with a duster or with blotting-paper. Put the ice in the beaker and again weigh; the increase in weight will give the weight of ice added. When the ice is all converted into water, the resulting temperature, T_3, is noted.

Then if x denote the latent heat of water, nx is the number of units of heat required to convert n grams of ice into water at 0° C.

Also nT_3 is the number required to raise n grams of water from 0° to T_3.

∴ Total heat gained by ice = $nx + nT_3$.

Heat lost by water = $W_2(T_2 - T_3)$; and assuming these equal, then $nx + nT_3 = W_2(T_2 - T_3)$, from which x is easily found.

EXAMPLE.—Into a beaker containing 200 grains of water at 40° C. 60 grains of ice are placed. When all the ice is melted, the resulting temperature is found to be 13° C.; find the latent heat of water.

Let x denote the latent heat; then $60x$ units of heat are required to convert the ice into water at 0° C.

SPECIFIC HEAT—CHANGE OF STATE. 173

Also $60 \times 13 = 780$ units from $0°$ to $13°$.
But the water loses $200 (40 - 13) = 5400$.
∴ $60x + 780 = 5400$.
$$x = \frac{5400 - 780}{60} = \frac{462}{6}$$
$$= 77 \text{ units}.$$

There are many sources of error in the experiment, and the value obtained is too small: thus the radiation of heat from the beaker and the heat given up by it in cooling from $40°$ to $13°$, together with any water carried into the beaker with the ice—all these make x too small.

The exact value obtained by the most careful experiment for *the latent heat of water*, or the amount of heat required to convert unit mass (1 gm. or 1 lb.) of ice at $0°$ C. into water at the same temperature, is 79 units. In a similar manner, when water freezes there are 79 units of heat given up by every unit mass. During this solidification the volume is increased, but the temperature is unaltered.

APPARATUS.—Glass tubing; beaker; freezing mixture.

EXPERIMENT 123.—Blow a bulb on one end of a glass tube and draw the other end out to a fine point. Fill the tube and bulb with water, and seal up the end of the tube. Place the bulb in a freezing mixture of ice and salt. The water in the bulb will be converted into ice, but owing to the expansion of the ice the bulb will be broken.

EXPERIMENT 124.—Dissolve some common salt in a beaker, and carefully note that the thermometer during the operation indicates that the water becomes colder, owing to the heat which is required for solution being taken from the water.

When common salt and ice or snow are used to form a freezing mixture, as both materials are solids, there is a supply of heat demanded to change their state; hence the temperature falls below $0°$ C.

Latent Heat of Steam.

APPARATUS.—A tin or glass flask in which steam can be generated, called a calorimeter; a piece of glass tubing bent twice at right angles as shown in Fig. 144, one end of which is inserted in a piece of straight tube, the other passing through a

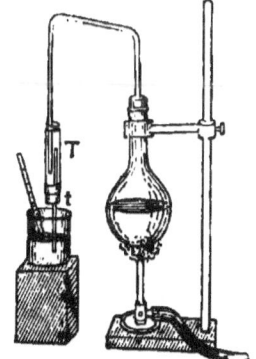

FIG. 144.—To ascertain the latent heat of steam.

SPECIFIC HEAT—CHANGE OF STATE.

tightly-fitting cork in the flask. A straight piece of tubing t passes through a cork at the lower end of the larger tube T, and projecting some little distance into the tube as shown, prevents water from passing into the beaker with the steam.

EXPERIMENT 125.—Having ascertained the weight of the beaker, a known quantity of water of weight W_w is placed in the beaker, and its temperature, T_1, noted. Steam is passed into the water, raising its temperature to T_2. Let W_s denote the increase in the weight (or weight of condensed steam), which is obtained by weighing beaker and contents after condensation. If W_b denote the weight of the beaker, the heat required to raise the temperature of the beaker from T_1 to $T_2 = W_b \times \cdot 2(T_2 - T_1)$, since the specific heat of glass is $\cdot 2$.

Heat units given to water $= W_w(T_2 - T_1)$.

∴ Heat units given to water and beaker $= (W_w + \cdot 2W_b)(T_2 - T_1)$.

Each gram of condensed steam has given up its latent heat, L, and the temperature has fallen from 100° to T_2; neglecting losses by radiation, etc.

Heat gained by the water and beaker = heat lost by the steam.

$$(W_w + \cdot 2W_b)(T_2 - T_1) = W_s[L + (100 - T_2)].$$

L, from this, is found to be nearly 536; or, ∴ unit mass (1 gm. or 1 lb.) of steam, in condensing to water at the same temperature, gives out 536 units of heat.

As in the case of finding the latent heat of water, there are several sources of error in the experiment—thus the losses by radiation and conduction have been neglected, also any *water* carried over by the steam—and all combine to make the value of L *too small*.

Repeat the experiment, altering the weight of the water.

Tabulate as follows:—

Weight of beaker..	=	grams.
Weight of beaker and water..................................	=	,,
Temperature of water before passing in steam......	=	,,
Temperature of water after passing in steam........	=	,,
Weight of steam condensed..................................	=	,,

EXAMPLE.—Steam is passed into a beaker containing 190 grams of water at a temperature of 17° C., the water being raised from 17° to 44°, or through 27° C. The increase in weight is found to be 9 grams, the weight of the beaker 30 grams. Find the *latent heat of the steam*.

SPECIFIC HEAT—CHANGE OF STATE.   175

Weight of beaker = 30 grams.
Weight of beaker and water = 220 grams.
Temperature of water before = 17°.
Temperature of water after = 44°.
Weight of steam condensed = 9 grams.
Heat given to water = $190(44 - 17) = 190 \times 27 = 5130$.
Heat given to beaker = $30 \times \cdot 2 \times 27 = 162$.
Heat given up by steam = $9(L + 100 - 44) = 9L + 504$.
∴ $5130 + 162 - 504 = 9L$.
∴ $L = \dfrac{4788}{9} = 532$.

(The accurate value of L is 536.)

SUMMARY.

Capacity for Heat.—Different substances have in general unequal capacities for heat.

Quantity of Heat.—When equal masses of different substances are heated or cooled through the same range of temperature, the quantities of heat absorbed or given out are in general different.

Specific Heat.—The specific heat of any substance is the (quantity, or) number of units of heat required to raise the temperature of unit mass (1 gram) through 1° C.

Change of State.—At certain definite temperatures many solids are found to change into liquids; also liquids are found to change into gases under similar conditions. Change of state is called *fusion*, and after fusion begins no change of temperature occurs until fusion is complete.

Latent Heat.—**The amount of heat required to change the state of a body, and which is not indicated by a thermometer—that is, without changing its temperature—is called latent heat.**

EXERCISES.

1. If 1 lb. of boiling water is mixed with 3 lbs. of ice-cold mercury, what will be the temperature of the mixture?

2. 500 grams of metal are heated to 99·5° C. and placed in 450 grams of water at 15° C.: the temperature rises to 23·5° C.; find the specific heat of the metal.

3. If 125 grams of a substance which has been heated to 140° C. are dropped into 193 grams of water contained in an iron calorimeter which weighs 63 grams, the temperature of the water rises from 15° C. to 20° C.; if the specific heat of iron is $\frac{1}{8}$, what is the specific heat of the substance?

4. The specific heat of lead is 0·031, and its latent heat 5·07; find the amount of heat necessary to raise 15 lbs. of lead from a temperature of 115° C. to its melting-point, 320° C., and to melt it.

5. Explain what is meant by the statement that the latent heat of water is 79.

If 10 grams of ice at the freezing-point be put into 100 grams of water at 18° C., what will be the temperature when all the ice has melted?

6. 200 grams of ice are placed in 100 grams of water at 100° C.: the temperature falls to 70°; find the latent heat of ice.

7. The weight of water in a beaker is 7,000 grains at a temperature 65° F.; steam is passed in until the temperature becomes 126° F.; the increase in weight is found to be 416 grains: find the *latent heat* of steam.

CHAPTER XVIII.

TRANSMISSION OF HEAT: CONDUCTION—CONVECTION—RADIATION.

Transmission of Heat.—Heat is transmitted by three processes, called **conduction, convection,** and **radiation.**

Conduction—Solids.

APPARATUS.—Wires 3 or 4 inches long, copper, brass, iron, and platinum; also glass tube and splinters of wood.

EXPERIMENT 126.—If the end of the copper or brass wire be held in the Bunsen flame for a very short time, it becomes too hot to hold; a piece of iron can be held for a longer time, a piece of glass longer than the iron; a splinter of wood can be held until the flame burns the wood almost close to the fingers—showing the difference which exists in different solids for conducting heat.

APPARATUS.—A piece of brass tubing, about 2 or 3 inches long, has a piece of wood fitted into it so as to form a round rod, half wood, half brass, about 3 or 4 inches long.

EXPERIMENT 127.—Wrap a piece of paper round the rod and hold it for some time in a Bunsen flame; the paper round the brass will not be scorched readily, but the portion round the wood is burnt. The brass being a good conductor of heat, the heat from the flame is carried rapidly away; but when wrapped round the wood, which is a bad conductor, the paper is scorched.

APPARATUS.—Ingenhausz apparatus.

EXPERIMENT 128.—We may compare the thermal conductivity of metal and other rods by the Ingenhausz method.

The apparatus consists of a metal trough having a number of small tubes in one side, into which, fixed in with corks, rods of different materials can be placed. (Fig. 145.)

TRANSMISSION OF HEAT: CONDUCTION.

The rods are of equal lengths, about 10 or 12 cm., and 4 or 5 mm. diameter, and project equal distances into the trough. They are evenly coated with a covering of wax. The trough is then filled with boiling water, and the melting of the wax shows the progress of the heat along the rods. When there is no further sign of melting, the *distance* to which the melting extends represents the conductivity.

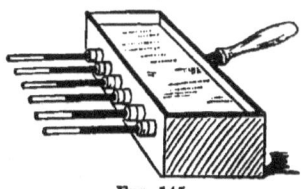

FIG. 145.
Ingenhausz apparatus.

Make, if possible, two or three trials, and tabulate the results.

APPARATUS.—Two strips or bars of the same size, and of iron and copper respectively.

EXPERIMENT 129.—Small wooden balls, as shown in Fig. 146,

FIG. 146.—Experiment to show the conductivities of iron and copper.

are attached to the strips by means of wax, when fixed end to end as shown; or better, riveted to a cross-piece (the cross-piece being heated), and a flame placed underneath. The balls are found to fall away from the copper much more rapidly than from the iron.

EXPERIMENT 130.—If a piece of gauze be held over a gas burner when the gas is turned on, the gas may be ignited, and

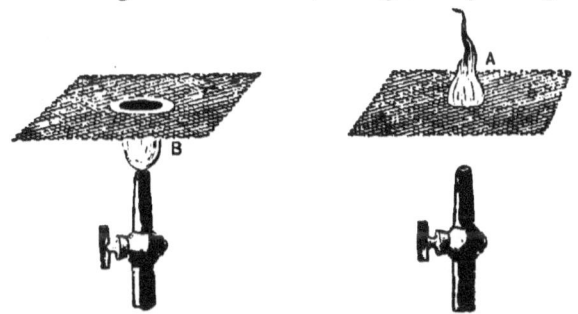

FIG. 147.—Action of wire-gauze on a flame.

will burn above the gauze (as shown at A, Fig. 147), but does not catch fire underneath. The gauze being a good conductor of heat, rapidly conveys the heat away from the portion where the flame is, and thus the gas underneath is not ignited. If the gauze be lowered on to the flame, the flame appears to be pressed down by it as shown at B, and does not light above it until the gauze has become heated to whiteness.

Fig. 148. Davy safety lamp.

Davy Safety Lamp.—This simple principle is usefully employed in the *Davy safety lamp* (Fig. 148). The gauze is a good conductor of heat, and the heat from the lamp flame is diffused over the whole extent of the gauze, being so reduced in heating power as to be unable to set fire to any explosive mixture, such as "fire-damp," which may surround the surface of the lamp.

Conduction—Liquids.

APPARATUS.—A long test tube, water, piece of ice, and Bunsen burner.

EXPERIMENT 131.—Put the piece of ice at the bottom of the test tube; the ice must be fastened to a piece of lead, to prevent it rising when water is added above. Fasten the tube in an inclined position by means of a retort stand. Boil the water in the upper part of the tube by means of the Bunsen flame, and note that the ice does not melt. The experiment will show that it takes a comparatively long time for the ice to melt; hence water is a bad conductor of heat. It will be found that all liquids, with the exception of mercury, are bad conductors.

Conduction—Gases.

EXPERIMENT 132.—Place a little powdered lime in one hand. Place on this the point of a piece of iron, such as a poker, heated to redness. There is a good deal of air present with the lime, and as lime is a bad conductor, if any heat is felt by the hand when a hot poker is made to rest on the lime, it must be conducted to the hand by the air. It will be found that heat does not travel to the hand; hence *air is a bad conductor*. This may also be shown if a poker be made red-hot, and the shadow which it casts on a wall or screen be observed. The wavy motion due to the heated air surrounding is only discernible close to the shadow; hence air is a bad conductor.

Convection.

EXPERIMENT 133.—Heat over a small Bunsen flame or a lamp (as shown in Fig. 149) a round-bottomed flask full of water, and drop into the water some colouring matter, such as crystals of magenta dye, or aniline dye. The heated water, as shown by the arrows, ascends, and the colder water from the sides rushes in to fill the place. These convection currents are easily observed by the movements of the colouring matter referred to.

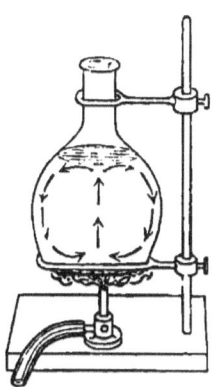

FIG. 149.
Convection currents.

The system of heating buildings by hot-water pipes affords an excellent illustration of all the three processes of *conduction, convection,* and *radiation.* The system consists of a boiler situated generally at or near the basement of a building. A set of pipes passes from the top of the boiler, through the various rooms to be heated, and back again to the bottom of the boiler.

The water above the furnace in the boiler, being heated by conduction, is caused to expand, and thus becomes lighter (bulk for bulk) than when it was at a lower temperature; it therefore rises up the pipe leading from the top of the boiler, its place being taken by the colder water entering the pipe at the bottom of the boiler. This in turn is heated, and rises. As the hot water circulates through the pipes, it communicates most of its heat to the pipes by conduction, and these in turn radiate the heat in all directions, heating the air by *convection and radiation.*

SUMMARY.

Transmission of Heat.—Heat is transmitted by the three processes of *conduction, convection,* and *radiation.*

Conduction is the name given to the passing of heat from particle to particle of a body.

Solids are often divided into good and bad conductors. Metals are good conductors; wool, sand, ivory, flannel, are bad conductors.

Liquids and gases are in general bad conductors. In the *Davy safety lamp,* the metal gauze being a good conductor, quickly carries heat away from any portion.

EXERCISES.

1. A layer of oil six inches deep floats on water of the same depth. Describe the movements of the liquids when the water is gently heated from below.

2. The bulb of a delicate thermometer is immersed at a slight depth below the surface of water. The upper surface of the water is heated, and the thermometer is hardly affected. On heating the water from below, the thermometer shows an immediate rise of temperature. Explain these two results.

3. Why is a block of ice so slow in melting even in a warm room? What are the circumstances which regulate the rate of evaporation of a liquid?

CHAPTER XIX.

MAGNETISM.

Properties of Magnets—Magnetic Induction.—There are certain hard black stones found in various parts of the world which possess the property of attracting iron and steel. Small pieces of iron and steel, such as filings, etc., when one of these stones is brought near to them, are attracted, and cling to the stone. The name **lodestones** was given to these stones. Also, as they were found at a place called Magnesia, the name Magnes or **Magnet** was given to them. When carefully examined, they are found to consist of iron and oxygen in the proportion of 64 parts by weight of oxygen to 168 parts of iron, the formula being Fe_3O_4. If a piece of iron or steel be rubbed with one of these stones, the magnetism is imparted to the iron or the steel: the former are called **natural magnets**, the latter **artificial magnets**. *Steel magnets* are made of various forms—the more common being a rectangular strip of steel, called a *bar magnet*, and a piece of steel bent into the form of a horse-shoe, and called a *horse-shoe magnet*. **Artificial magnets** are either *temporary* or *permanent*—the former of soft iron, and the latter of steel. In the former case, their magnetic condition is due either to the presence of a permanent magnet, or, as in an electro-magnet, to the passage of an electric current. In a steel and permanent magnet, the magnetism is produced either by stroking with a magnet or by means of an electric current.

APPARATUS.—Bar magnet.

EXPERIMENT 134.—Present one end of the magnet to a small piece of iron or steel (a tin tack or a needle will do): it will be attracted to and adhere to the magnet. If one end or the whole length of the magnet be wrapped in a piece of paper or thin cardboard, the piece of iron or steel will be attracted as before—showing that the presence of the paper does not affect the attractive force.

MAGNETISM. 181

If a needle be placed on a sheet of cardboard, it will be found to move about and follow the motion of a magnet placed underneath the cardboard. Ascertain by experiment that the magnet will attract iron, steel, nickel, and cobalt, but not other metals, or such bodies as wood, stone, etc.

APPARATUS.—Bar magnets; iron filings.

EXPERIMENT 135.—Plunge the magnet in a heap of iron filings, or scatter filings over it, and notice the accumulations in the form of *tufts* at both ends of the bar, but that no filings adhere at the middle of the bar.

If the experiment be made carefully, the two tufts will be found to be alike—indicating that the attractive forces exerted by the bar at its ends are equal, also that the force (or magnetic force) resides chiefly at or near the ends of the bar. The two opposite points or ends of a magnet are called its *poles*.

If two magnets, after being dipped into the iron filings, be brought near to each other, then when two poles are presented, the tufts will approach, as shown at A (Fig. 150); if one magnet be reversed and the two poles are presented, the tufts recede, as shown at B.

FIG 150.—Magnets attracting iron filings; unlike and like poles.

APPARATUS.—Bar magnet; small pieces of iron, such as soft iron tacks.

EXPERIMENT 136.—Bring small pieces of soft iron, such as soft iron tacks, near to the ends of the magnet (Fig. 151), and it will

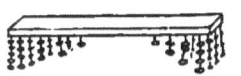

FIG. 151.—Magnet and small iron tacks.

be found that several tacks may be put on at certain points (in vertical rows), but the number of tacks supported at any point of the magnet will be found to depend on the distance of the point from the centre; at or near the centre no tacks are supported, and from this it would appear that the bar has no magnetism at the centre. But, as shown later in Experiment 144, if the bar be broken at the centre, the two ends so formed would attract pieces of soft iron as before.

As both ends of the bar magnet equally attract a piece of iron or steel, the advantage due to the use of the horse-shoe form, in which the two ends are brought together, is at once apparent.

In Fig. 152 a horse-shoe magnet and its keeper are shown. The effect of the magnet on the keeper AB will be to make it a temporary magnet, the end A a south pole, and the end B a north pole. The keeper being a magnet, and its polarity being opposite to that of the magnet against which it is placed, the two magnets exert all their force in attracting each other; thus there is very little free magnetism left to attract any outside body, such as a piece of soft iron held near it. This is described best by saying that the magnetic field is closed.

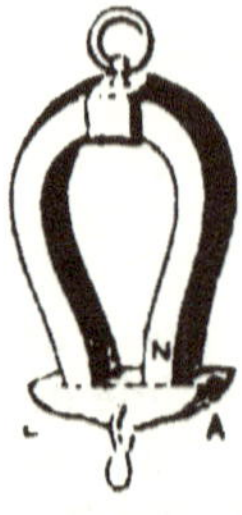

APPARATUS.—Horse-shoe magnet and keeper; various weights.

FIG. 152. Horse-shoe magnet and keeper.

EXPERIMENT 137.—Fix a hook to the keeper AB. It will be found on attaching weights that the supporting force exerted by the keeper when placed across both ends of the magnets is more than twice as great as when placed only on one.

APPARATUS.—Horse-shoe magnet; iron filings; sheet of cardboard.

EXPERIMENT 138.—Place a piece of cardboard over the horse-shoe magnet lying on the table, and scatter iron filings over it. When the cardboard is tapped, the arrangement of the filings is clearly seen (Fig. 153) to consist of a large number of symmetrical curves called *magnetic curves* or *lines of forces* (as they represent the lines of force round the magnet in a horizontal plane). The same arrangement exists in every possible plane. In this manner we are said to map the field of a magnet. If the *magnetic field* be closed by joining the poles of the magnet by an iron keeper, then the circuit is completed through the keeper rather than through the air. It will be found that the arrangement of the iron filings indicates only few *lines of force* beyond the iron keeper; the strong lines of force which spread out from the poles are concentrated in the keeper.

FIG. 153.—Horse-shoe magnet and magnetic curves.

APPARATUS.—Bar and horse-shoe magnets.

EXPERIMENT 139.—Suspend the bar magnet by a paper stirrup in such a manner that it can turn quite freely in a horizontal plane, and the magnet comes to rest in a direction north and south (magnetic). If the magnet be displaced, it will oscillate backwards and forwards, but will ultimately come to rest in the same direction as before.

If the horse-shoe magnet be brought near to the suspended magnet, it will be found that like-named poles repel, but the two unlike-named poles attract, each other. The end of the magnet pointing in the direction of the north is called the north or positive end; the opposite end is called the south or negative end.

Induction.

APPARATUS.—Magnets as before; small pieces of iron and steel.

EXPERIMENT 140.—Place the horse-shoe magnet in a vertical position, resting on its bend so that the two poles are upward. Hang by means of a silk string a small soft iron nail in a horizontal position over the poles: the nail will become a magnet. On testing its polarity by means of the pole of another magnet, repulsion will be obtained, if the pole of the movable magnet is of the opposite kind to that of the fixed one; or the end of the nail which is over the north pole of the fixed magnet is a south pole.

Remove the fixed magnet, and the nail ceases to be a magnet; or reverse the poles of the magnet, and a reversal of the polarity of the nail takes place. Raise the nail higher and higher, and observe that the effect of repulsion diminishes, and finally that the nail is no longer a magnet, but that both ends of the nail are attracted by the small or testing magnet.

EXPERIMENT 141.—Use instead of the soft iron a piece of hard steel, such as a needle. The steel at first is only a feeble magnet, but becomes stronger if left for some time; also, when the magnet is removed the needle remains a magnet. Place the magnet underneath with the poles reversed; the needle will turn itself round so that the north pole of the needle is over the south pole of the magnet.

Thus *soft iron* under the influence of a magnet is itself *easily magnetized*, but soon loses its magnetism when the magnet is removed; *steel* acquires it *with difficulty*, but retains it *permanently*.

Magnetizing Iron and Steel.

APPARATUS.—Bar or horse-shoe magnet; pieces of steel.

EXPERIMENT 142.—Lay a piece of soft iron on a table (Fig. 154), and stroke it several times (in the *same direction*)

FIG. 154.—Magnetizing a piece of iron or steel.

with one pole of a bar or horse-shoe magnet: the soft iron becomes a magnet for a short time, but very rapidly loses all its magnetism.

Replace the iron by a needle or a piece of unmagnetized steel ribbon or clock spring, and holding a bar or horse-shoe magnet in a nearly vertical position (Fig. 155), draw along it from one end to the other repeatedly one pole of the magnet; the rubbing must *not* be backwards and forwards, but always in the same direction. It will be found that the end last touched by the pole of the magnet has opposite polarity to that pole. Thus, if the piece of steel be rubbed with the north end of the magnet as shown, the direction of rubbing from left to right, then the end last touched—that is, the right-hand end—possesses south polarity.

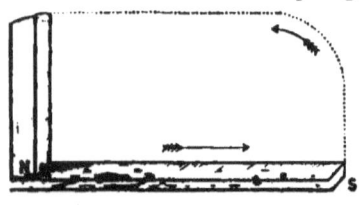

FIG. 155.—Magnetizing a piece of steel spring.

To test this, suspend the strip, and determine which end is the north pole.

APPARATUS.—Pieces of clock spring; magnets.

EXPERIMENT 143.—Fasten a piece of clock spring to the table, and rub it from the middle to the ends with the north-seeking poles of two magnets; show that the two ends of the strip are both south-seeking poles. Place the strip beneath a sheet of white paper on which filings are scattered, and notice that the consequent pole in the middle is north-seeking.

EXPERIMENT 144.—Heat a few inches of steel spring red-hot, and quench in water rapidly: this makes the steel very hard and brittle. Magnetize it as in Experiment 142 (Fig. 155), and mark the poles; break it in halves (marking the ends so as to know which were next to each other), quarters, etc. If each broken piece be tested, it will be found to be a permanent magnet, having polarity like the original.

APPARATUS.—Bar magnet; long and short strips of steel; test tube and steel filings.

EXPERIMENT 145.—Magnetize a long and a short strip, using the same number of strokes in each case. By bringing both near to the magnetic needle, observe that the longer piece has much the stronger polarity. Explain from this why it is difficult to magnetize a short, thick piece of steel. Show how two opposite poles which are in contact neutralize each other's effects, by putting the two parts of the broken strip in the position they occupied before the strip was broken; then plunge into filings.

MAGNETISM.

EXPERIMENT 146.—Fill a test tube with steel filings, and magnetize the filings. This may be done, as before, by drawing one end of a magnet several times along the tube. Suspend the test tube, taking care not to disturb the filings, and test the ends: they will be found to be north and south respectively.

If the tube be shaken so as to disturb the filings, it will be found that it is no longer magnetic, and is attracted by either end of the magnet.

SUMMARY.

A magnet is a body having the power of attracting iron and steel.

Natural magnets or **lodestones** are found at various parts of the earth's surface. The stones consist of iron and oxygen, the chemical formula being Fe_3O_4, or 64 parts by weight of oxygen to 168 parts by weight of iron.

Artificial magnets are either **temporary** or **permanent**. The former are of soft iron, the latter of steel. The magnetism in each case is produced either by stroking or by the electric current.

Steel slowly acquires magnetism, but retains the magnetism for a considerable time.

Soft iron is easily magnetized, but loses its magnetism quickly.

Poles.—The attraction is greatest at the two ends of a magnet. These are called the north and south poles respectively. At or near the middle the attraction ceases. *Unlike poles of two magnets attract each other; like poles repel.*

Lines of Force.—By putting over a magnet a sheet of cardboard, and sprinkling over it iron filings, the directions of the lines of force are seen.

Induction.—The north pole of a magnet induces a south pole in a piece of iron or steel.

A magnet suspended so that it can turn freely in a horizontal plane always sets in the same direction, one end pointing north, the other south.

EXERCISES.

1. A piece of soft iron and a piece of hard steel, of the same size and shape, are separately rubbed from end to end by the north pole of a strong bar magnet. How will you test their magnetic condition, and what difference will you find between them?

2. A bar magnet has consequent poles at its middle point. Describe its effect on the direction of a freely-suspended magnetic needle which is moved from one end of the bar magnet to the other at a constant small distance above it. The effect of the earth's magnetism on the needle may be neglected.

3. A piece of soft iron, placed in contact with both poles of a horse-shoe magnet at the same time, is held to it with more than twice the force with which it would be held if it were in contact with only one pole of the same magnet. Why is this?

4. A piece of cardboard is placed over a horse-shoe magnet lying on a table, and iron filings are scattered over it. Draw a diagram showing the arrangement assumed by the filings when the cardboard is tapped. What differences would be observed in the arrangement of the filings when the ends of the magnet were joined in turn by bars of (1) steel, (2) soft iron, (3) copper?

5. A glass tube is nearly filled with steel filings, and corked at both ends. It can then be magnetized by any of the ordinary methods, but loses its magnetic properties when shaken. Explain this.

CHAPTER XX.

TERRESTRIAL MAGNETISM.

Terrestrial Magnetism.—It has been found in the preceding experiments, that when a magnet is so suspended that it can turn freely in a horizontal plane, such as a magnetic needle shown in Fig. 156, it always points in the same direction, which is nearly north and south. It is found, however, that a magnet does not point exactly to the true geographical north, or the north end of the axis on which the earth turns.

Fig. 156.—Magnetic needle.

EXPERIMENT 147.—Suspend small magnet a short distance above a table. If underneath a strong magnet be placed in any direction, it will be found that the small magnet will turn so that its south end is directly over the north end of the larger one; and if the larger magnet be turned in any direction, the smaller one will always follow it: hence when a large magnet is brought near a small one, the smaller is controlled by the larger. Thus the suspended magnet, the magnetic needle (or the compass needle), always points to the north, controlled by *the earth, which behaves as a large magnet*, having its poles near to the geographical poles. But in the above experiment we have found that the south pole of the smaller magnet is directly over the north pole of the larger one; so that if the north pole of the earth be called north, what we called the north pole of a magnet is really a south pole. To avoid difficulty, the pole of a magnet which points to the north is usually called the *north-seeking pole*.

TERRESTRIAL MAGNETISM.

When accurate observations are made, it is found that a freely-suspended magnet does not point exactly due north: the angle between the direction of such a magnet and the true north is called **the declination**. The declination changes from point to point on the earth's surface, and also with time. For Great Britain the value at the present day is about $16\frac{1}{2}°$ W.; or the needle, instead of pointing due north, points to the west of it.

Inclination or Dip.

APPARATUS.—Magnetic needle; magnet; a few unmagnetized knitting-needles.

EXPERIMENT 148.—Support a freely-suspended magnetic needle above a table, and a few inches from a magnet placed on the table: the south pole of the needle will dip vertically downwards. If the needle be carried in a horizontal direction over the magnet, the inclination or dip will lessen until, when it reaches the centre, it will be horizontal; after passing the centre it again begins to dip, but in the opposite direction—that is, the north pole will point vertically downwards.

EXPERIMENT 149.—Fasten a silk fibre firmly with hot shellac to the middle of an unmagnetized knitting-needle. Suspend the needle, filing one end until the needle hangs horizontally. Magnetize the needle, and it will be found that it is no longer horizontal but inclined, with the north end pointing downwards.

The needle may be mounted on a horizontal axle so that it can turn freely in a vertical plane. Then in the northern hemisphere the north pole points downwards; this increases as the needle is carried north, until, if the needle is over the magnetic north pole, it points vertically downwards. When carried towards the equator, its inclination becomes less and less, until it is half-way between the north and south poles, when it is horizontal. When carried further south, the south pole of the needle points downwards.

When the needle is placed in the magnetic meridian, the angle which the needle makes with the horizontal line through its centre is called the inclination or dip. The straight line joining the north and south poles of a magnet is called the *magnetic axis*. The *magnetic meridian* may be defined as an imaginary line drawn from either magnetic pole through the axis of a magnetic needle at the place of observation.

APPARATUS.—Unmagnetized steel (knitting-needle), about 4 or 5 inches long; bar magnet.

EXPERIMENT 150.—Suspend the piece of steel in a horizontal position by a silk fibre. Magnetize it as carefully as possible. Replace, and note that it sets in the magnetic meridian, with the north-seeking end pointing in a downward direction. The angle which it makes with the horizontal (a little greater than 67°) is called the **angle of dip**.

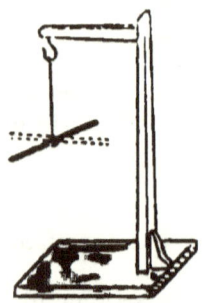

FIG. 157.—To show the angle of dip.

Magnetic Moments.—We have already found that the strength of a magnet depends upon its length. In a comparatively long magnet, the poles, being further apart, do not tend to neutralize each other's effect to the same extent as in a shorter one. Hence the strength of a magnet depends upon the strength of its poles and the distance between them; *the product of the strength of either of the two poles of a magnet, into the distance between them*, is called the *magnetic moment of the magnet.*

To compare Magnetic Moments.

APPARATUS.—Bar magnets; magnetic needle.

EXPERIMENT 151.—Set the suspended magnetic needle oscillating, placing a bar magnet at a distance of a few inches to the north of the needle. Count the number of vibrations per minute. Replace the magnet by another of, as nearly as possible, the same strength, and at the same distance, and again count the oscillations. Shift the magnet to an equal distance south of the needle, the north-seeking end being nearest the needle. The number of oscillations will be found to be the same as before. Replace the first magnet to the north of the needle. It will be found that the number of vibrations, under the influence of both magnets, is considerably less than the sum of the numbers of the vibrations due to each magnet separately—the magnets being of nearly equal strength, and the earth's field being neglected. More exactly, the number of vibrations with the two magnets is 82, or 1·4 times as great as when only one is used. By means of the above, verified by other and more elaborate experiments, it is found that *the vibrations of a magnet in a given time are proportional to the square root of the strength of the magnetic field in which the magnet swings.*

If a steel magnetized needle be freely suspended, and two pieces of lead perforated so that they can be placed at any point on the needle, it is easily shown, by placing the pieces at any

TERRESTRIAL MAGNETISM.

convenient equal distances from the centre of the needle and counting the vibrations in a given time, that the *time of vibration depends upon the mass and the shape of the vibrating body.*

The effect of the earth must be taken into account. If we denote by a the number of swings due to earth alone, by b the number of swings due to earth and the first magnet, by c the number of swings due to earth and the second magnet, we have the relation—

$$(b^2 - a^2) : (c^2 - a^2) : : \text{strength of A} : \text{strength of B}.$$

Time of Vibration of Magnet.

APPARATUS.—Number of steel needles; bar magnet; small cylinders of lead.

The time of vibration of a magnet depends upon the mass and the shape.

EXPERIMENT 152.—Suspend a steel knitting-needle; take two small cylinders of lead, and bore holes through them so that the needle will pass through. Magnetize the needle and suspend it as before. By means of the movable lead cylinders the time of vibration can be ascertained for any position of the cylinders.

Time of Vibration depends on Magnetic Moment.

APPARATUS.—A number of unmagnetized knitting-needles; bar magnet.

EXPERIMENT 153.—Suspend a number of unmagnetized needles, and determine time of vibration. Magnetize one, two, three, etc., and in each case find the time of vibration.

Magnetizing Effect of the Earth.

APPARATUS.—A bar of soft iron (a poker); hammer.

EXPERIMENT 154.—Hold a poker in the magnetic meridian and in the magnetic dip—in other words, parallel to the magnetic axis of the earth. Hit the upper end with a hammer: on testing it with the compass, it will be found to have become magnetic in such a way that the north-seeking end is downwards and towards the north. This shows that the magnetism at the north pole of the earth is of the same kind as in the south-seeking end of a magnet. In other words, the earth, regarded as a large magnet, has its south pole near the geographical north pole, and its north pole near the geographical south pole.

Summary.

The Earth behaves as a large magnet in its action upon other magnets, its south-seeking pole being near the geographical north pole.

Declination.—A freely-suspended magnet is found not to point exactly due north: the angle between the direction of such a magnet and the true north is called the *declination*.

Inclination or Dip.—A magnetic needle, when balanced on its middle point, does not rest in a horizontal position: the angle between the direction of the needle and a line through its centre is called the *inclination* or *dip*.

Magnetic Moment.—The product of the strength of either of the two poles of a magnet and the distance between them is called the *magnetic moment* of the magnet.

Time of Vibration of a freely-suspended magnet depends on its magnetic moment, and also upon the mass and the shape of the vibrating body.

Exercises.

1. A steel bar suspended by a thread lies horizontally, and points indifferently in any direction; but when it is broken into halves, each half is found to point north and south when separately suspended like the whole bar. Explain the magnetic condition of the unbroken bar.

2. What is meant by saying that the magnetic dip at London is 67° 30′?

3. An iron rod held vertically is tapped with a mallet. The upper end is found to repel the south pole and attract the north pole of a compass needle. The rod is now quietly inverted, and the same end (which is now the lower) is tested again. It is then tapped and once more tested. State what results you would expect, and explain them.

4. It is suspected that a magnetized bar of steel has consequent poles. How would you ascertain whether this is so or not?

5. A rod of iron, when brought near to a compass needle, attracts one pole and repels the other. How will you ascertain whether its magnetism is permanent, or is due to temporary induction from the earth?

6. Given a magnet and the means of suspending it, how will you determine (1) the magnetic meridian; (2) in which direction *north* lies? It is assumed that you do not know which end of your magnet is a north and which a south pole.

CHAPTER XXI.

ELECTRIFICATION BY FRICTION—POSITIVE AND NEGATIVE ELECTRIFICATION—CONDUCTORS AND INSULATORS.

APPARATUS.—Sticks of ebonite and sealing-wax; flannel and silk rubbers; round-bottomed flask; lath of wood; piece of copper wire.

EXPERIMENT 155.—Rub a stick of ebonite smartly with the flannel or silk, and hold the excited rod near to small pieces of paper or other light objects, such as small chips of matchwood, bran, etc. These will be attracted to it: some will cling to the rod, and others will dance up and down between the rod and the table. The ebonite is said to be "electrified." As a magnet will only attract metals such as iron or steel, and the ebonite stick will attract all light bodies, this property is different from magnetism.

EXPERIMENT 156.—Balance the lath upon the bottom of the round-bottomed flask inverted. If the stick of rubbed ebonite be brought near to one end of the lath, it will be found to be attracted towards the ebonite. Conversely, if the stick of ebonite be laid on a suitable support so that it is free to move, and the lath or a pencil be brought near to it, the ebonite will be attracted, and will move towards it.

EXPERIMENT 157.—Rub an ebonite rod with flannel so that it becomes electrified, and place it in a wire stirrup made from a piece of copper wire, bent as shown in Fig. 158. In the previous experiments we have found that an electrified rod attracts light objects; it also attracts

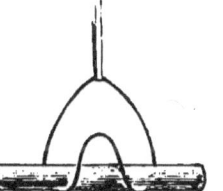

FIG. 158.—Wire stirrup.

comparatively heavy bodies (as in the case of the balanced lath), but the attractive force is not sufficient in amount to lift them. If the electrified rod be placed in the stirrup, and a pencil or lath of wood brought near, the electrified rod will be attracted towards it. Repeat the experiment, using rods of iron, brass, etc. Show that they all attract the electrified rod.

APPARATUS.—Glass rod; stick of shellac; flannel and silk rubbers.

EXPERIMENT 158.—Rub a stick of shellac with dry flannel, and show by its action on light substances (as in the case of the ebonite rod) that it is electrified.

Excite glass with amalgamed silk, also sealing-wax with flannel; place each in succession on a wire stirrup, as shown in Fig. 158. Note how each will move towards a finger held near.

The stirrup shown may be replaced by a paper stirrup.

Amalgamed silk is prepared by melting together zinc and tin in the proportions of one part tin to two parts zinc, and when nearly cold stirring in from seven to ten parts of mercury. The amalgam, when cold, can be made to adhere to the silk by using a small quantity of lard (freed from salt).

Positive and Negative Electrification.

EXPERIMENT 159.—Electrify a stick of ebonite and place it in the stirrup. Bring near to one end of the suspended rod a similarly electrified rod, and instead of attraction repulsion will be observed.

In a similar manner electrify a stick of sealing-wax; show that when a second electrified stick of sealing-wax is brought near to it, *repulsion* takes place.

Electrify a rod of glass by rubbing with amalgamed silk. Bring it near the stick of sealing-wax, and notice the *attraction*. Thus the electrification induced in the glass rod is different from that in the stick of sealing-wax. The glass rod is positively electrified, the sealing-wax negatively electrified.

In the above experiments we find that the electrification in glass when rubbed with silk differs from that in ebonite, sealing-wax, or shellac when rubbed with flannel.

The repulsions and attractions are briefly expressed as—

1. All bodies which are electrified attract all unelectrified bodies.
2. Like electricities repel each other.
3. Unlike electricities attract each other.

Electrification by Contact.

APPARATUS.—Pith ball; unspun silk; sticks of ebonite and sealing-wax; flannel and silk rubbers.

EXPERIMENT 160.—Hang up the pith ball by means of the silk thread. Electrify a rod of ebonite and present it to the pith ball, which will be attracted to it until the rod touches the ball, when it is repelled by it, but is attracted when a glass rod rubbed with silk is presented to it.

It will be found that the ball will now attract, or can be attracted by an unelectrified body such as the hand. Instead of the pith ball, the end of the silk thread may be used; hence the silk does not conduct away the electrification.

EXPERIMENT 161.—Suspend the pith ball by a piece of *cotton thread*. Bring an electrified rod near to it: it will be attracted, but when touched by the rod it falls away. On bringing an unelectrified body near it, it is found that no attraction takes place. All the charge of electricity given to it by the rod has disappeared. This was not the case when the ball was suspended by the silk thread; hence we infer that *cotton* is a *conductor*, but silk is a *non-conductor*.

SUMMARY.

Electricity can be induced by rubbing ebonite, sealing-wax, etc., with flannel and silk rubbers. The electricity is shown by the attraction of light bodies.

Electrified bodies are found to attract all bodies *unelectrified*.

Like *electricities* **repel, unlike attract** each other.

Insulators or *non-conductors* are bodies—such as silk, glass, shellac, etc.—which allow little electricity to pass through them.

Conductors readily allow electricity to pass through them, and are divided into *good conductors*—metals, water, the earth; and *poor conductors*—cotton and wood.

The best conductors are found to offer some resistance to the passage of a current; also the best insulators permit a certain amount of electricity to escape.

EXERCISES.

1. A rod of sealing-wax and a piece of flannel, after having been rubbed together, are insulated and placed some distance apart; how do their potentials differ from each other and from the potential of the earth? How would you prove the truth of your answer?

2. A pith ball is suspended from a metal stand by a fine thread. If you have a strongly-electrified glass rod, how can you find out whether the thread is a conductor or a non-conductor of electricity?

3. Describe any experiment by which you could prove that when electrification of one kind is produced, the opposite kind is also produced in equal quantity.

4. Two pairs of light pith balls are hung at the opposite ends of an insulated conductor—one pair being suspended by silk, and the other by cotton threads. Describe and explain the behaviour of the balls if the conductor is gradually electrified more and more strongly.

(1,004)

CHAPTER XXII.

ELECTROSCOPE—ELECTRIC INDUCTION—PROOF PLANES—ELECTROPHORUS.

Gold-Leaf Electroscope.

APPARATUS.—Two strips of gold leaf, about 1 inch × ¼ inch, are fastened one on each side of the flattened end of a rod of brass.

Fig. 159.—Gold-leaf electroscope.

At the upper end of the rod a brass disc, or a knob, as shown in Fig. 159, is fixed; the rod passes through a cork, which is covered with sealing-wax to insulate the rod. Or a rod of ebonite, passing through an india-rubber stopper, may be used. To the lower end the strips of gold leaf, called "leaves," are attached, and the top is fitted with a brass disc, called the *cap* of the electroscope; a small hole in the disc allows wires to be attached to it.

The flask should be carefully cleaned and dried. The glass flask prevents currents of air from affecting the leaves, and also ensures that they are not in contact with other bodies.

EXPERIMENT 162.—*To charge the electroscope* positively, bring near to it an electrified glass rod, and the leaves will diverge; touch the disc with the rod, and when the rod is in contact put the cap to earth by touching it momentarily with the finger; when the rod is withdrawn the leaves will diverge, showing that the instrument is charged.

In a similar manner the electroscope may be charged negatively by using a rod of sealing-wax or ebonite rubbed with flannel.

EXPERIMENT 163.—*To ascertain the nature of the charge in an electrified body*, charge the electroscope as in previous experiment. Bring (1) a charged glass rod (positive), (2) a charged rod of sealing-wax (negative), near to the cap of the electroscope; notice whether the leaves diverge or collapse. In a similar manner any other charged body may be brought up to the cap of the electroscope, and the effect produced by it on the "leaves" (whether the same as or opposite to that produced by the glass rod) should be noted.

APPARATUS.—Gold-leaf electroscope; several yards of copper wire; silk and cotton thread; sealing-wax.

EXPERIMENT 164.—Fasten one end of a fine copper wire, 4 or 5 yards long, to the metal disc of the electroscope; make a loop at the other end and pass it over one end of a glass rod. Rub the other end of the glass rod with amalgamed silk. Although + electricity is developed on the glass, there is no divergence of the leaves of the electroscope—that is, the electricity at one end of the glass does not pass through the glass and wire to the electroscope. Thus the copper wire or the glass rod, or both, are bad conductors. Slide the loop, without touching it, along the glass rod until it approaches the electrified part; when this occurs the leaves diverge. Hence the copper wire is, but the glass rod is not, a conductor.

EXPERIMENT 165.—(*a*) Repeat the last experiment, using a stick of *sealing-wax* instead of the glass rod.
(*b*) Replace the copper wire by *cotton thread*.
(*c*) Use *silk thread* instead of cotton.
(*d*) Dip the silk thread in water, and note what alteration, if any, occurs.
(*e*) Replace the copper wire as in (*a*), place the rod on the ground or on a conductor, with a good earth contact, such as a metal pipe, connected with the earth.

EXPERIMENT 166.—Charge the electroscope as in Experiment 154. When the surrounding air is very dry, the leaves remain divergent for a considerable time. Thus *dry air is a bad conductor*. When the surrounding air is damp, the leaves soon collapse.

Hence it is easily seen that experiments in frictional electricity succeed better when the atmosphere is dry. In addition, all frictional apparatus should be perfectly dry. It is found that glass allows water to condense freely on its surface. To prevent this, a coat of shellac is applied to the glass.

Proof Planes.—It is often necessary to examine the electricity on a body, and to obtain a measure of the kind and quantity of the electrification, without manipulating the body itself; this may be effected by means of what is called a *proof plane*, which consists of a small disc of metal, fastened to the end of a rod of ebonite, sealing-wax, or varnished glass. A simple form is to fasten a penny to the end of a stick of sealing-wax, or to paste tinfoil or Dutch metal on cardboard discs of various sizes.

EXPERIMENT 167.—Suspend a small file by means of two silk threads; the file will thus be insulated. Bring near to one end a piece of rubbed ebonite, and touch the other end of the file with the proof plane. Remove the proof plane, and test the charge by means of a charged pith ball, as in Experiment 160: it will be found that it is negatively electrified.

Repeat the experiment, using a rod of rubbed glass, and therefore positively electrified, instead of the ebonite rod.

Induction.—We have found, in a preceding experiment (140), that when a piece of soft iron is placed near a magnet it becomes magnetized, the magnet having the power of acting upon the iron at a distance from it. The iron is said to be magnetized by *induction;* the magnetism in the iron is called *induced magnetism*.

A similar phenomenon occurs in the case of electrified bodies.

If a charged ebonite rod be brought near to an insulated and unelectrified conductor, the charge on the electrified rod is called the *inducing charge;* the charge developed, the *induced charge*—the conductor being electrified by *induction*.

APPARATUS.—Wooden rod, ebonite and glass rods, and rubbers.

EXPERIMENT 168.—Hold near the end of a long wooden or metal rod, placed on an insulated support, an electrified ebonite or glass rod. It will be found, as before, that electricity has been developed in the first-named rod, and that the opposite end of the rod becomes electrified, and will attract small substances, such as pieces of paper, etc.

EXPERIMENT 169.—Bring the charged rod near to an insulated conductor, and while in this position let the opposite end of the conductor be touched by the finger: this forms for a short time a connecting body with the earth. On removing the hand, the charged rod being kept in its place, the conductor will be found, by testing, to be charged all over with the opposite kind of electricity to that of the charged rod; and the charged rod being removed, the conductor will be permanently charged. Thus the effect of bringing the charged rod near to the insulated conductor has been to draw to the end nearest the rod a charge of the opposite sign, and the charge of the same sign is driven to the other end; this escapes when the earth connection is made. The portion which escapes is called the **free charge**; that which remains, and which is opposite to that of the charging body, is called the **bound charge**.

APPARATUS.—Gold-leaf electroscope, glass rod, and rubbers.

EXPERIMENT 170.—Bring near to the knob a charged rod; the leaves will diverge—due to the separation of the charges in the *rod* and *leaves*. From what we have already found in previous experiments, an unlike charge to that of the rod is attracted towards the knob, and a like charge is repelled to the leaves at the opposite end. If the knob be touched by the finger, the momentary contact puts the charge to the earth, and the leaves collapse: the charge is now of the opposite kind to that of the rod. On removing the charged rod the leaves again diverge, owing to the opposite charge, which was held very strongly near the knob, but faintly at the leaves, spreading itself over the whole. It should be noted that collapse of the leaves is not always an indication that the body which is being tested is charged; increased divergence is a much more reliable indication. Any unelectrified conductor, such as the hand, when brought near, would cause the leaves to collapse.

EXPERIMENT 171.—Charge the electroscope positively by a rubbed glass rod, negatively by an electrified rod of ebonite or sealing-wax. Make tests of the charge, and write out the tests you make in order to ascertain if the electricity is of the required sign.

Electrophorus.

APPARATUS.—A metal disc, B (Fig. 160), furnished with an insulating handle, G, of sealing-wax or glass; a disc, A, of insulating material, such as ebonite, shellac, or vulcanite. On the bottom surface of A a disc of tinfoil is pasted.

EXPERIMENT 172.—Rub the vulcanite lightly with fur; place the disc, B, on the vulcanite, and whilst in contact touch the disc with the finger. If the disc be now removed, it will be found that it has been charged by induction with the opposite kind of electricity, and if touched a spark can be obtained.

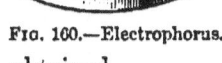

FIG. 160.—Electrophorus.

If the disc be again replaced on the vulcanite, it may be charged as before. The process may be repeated until all the charge on the vulcanite has disappeared.

ELECTROSCOPE—ELECTRIC INDUCTION.

Faraday's Ice-pail Electroscope.

APPARATUS.—Tin or copper vessel, 5 or 6 inches deep and 3 to 4 inches diameter; glass plate; electrophorus; electroscope.

EXPERIMENT 173.—Support the vessel on the glass plate to insulate it, and join the vessel to the disc of the electroscope by means of a fine metal wire. When a brass ball attached to a piece of silk string, which has been charged by means of the electrophorus, is lowered into the vessel without touching it, the leaves of the electroscope will be found to diverge, and the divergence will remain the same for any position of the sphere inside the vessel; if the sphere be allowed to touch the vessel, no alteration in the divergence of the leaves will take place. When the sphere is withdrawn, if it has not touched, the leaves of the electroscope collapse.

If, while the sphere is within, the vessel be momentarily put to earth by contact with the finger, the leaves collapse; when the sphere is withdrawn, the leaves diverge.

Write out a full account of the experiment, and state any deductions you can make from it.

SUMMARY.

Electroscope is an instrument used to detect the presence of electricity in a body.

Gold-leaf electroscope consists of two insulated leaves of gold leaf hanging near to each other. When a charge of electricity is given to them, they repel each other.

Electrophorus is an instrument for easily obtaining a charge of electricity, or for detecting the presence of electricity in a body.

Proof plane consists of a small, thin metal disc attached to an insulated handle, and may be used to compare the quantities of electricity on each equal area of a conductor.

Faraday's ice-pail electroscope is an instrument used to charge a body with any number of times the charge of a given body; also to prove that on a conductor equal quantities of positive and negative electricity are induced.

EXERCISES.

1. Say exactly what you must do in order to get a succession of sparks from an electrophorus.

2. State the disadvantages of glass as an insulator, and describe the best means of overcoming them.

3. To protect a gold-leaf electroscope from being acted on when an electrical machine is at work near it, it is sufficient to cover the electroscope with a thin cotton cloth. How is this?

4. The ebonite portion of an electrophorus is charged with electricity; what means would you take to completely discharge it?

5. A stick of sealing-wax having been rubbed with flannel, is found to be negatively electrified. How, by means of it, would you charge a proof plane with positive electricity?

6. An insulated conductor, A, is brought near to the cap of a gold-leaf electroscope which has been charged positively. State and explain what will happen (1) if A is unelectrified; (2) if it is charged positively; (3) if it is charged negatively.

CHAPTER XXIII.

DIFFERENCE OF POTENTIAL—ELECTRIC DENSITY.

Potential.—When two vessels contain water, one of which is at a higher level than the other, water will flow from the place of higher to that of lower level. In a similar manner, when two bodies having different temperatures are placed together, heat will pass from the hotter to the colder body, until both are at the same temperature. If a body having some electrical potential be brought into electric communication with a body having a different electrical potential, the two will as far as possible become of the same potential.

If the body having electric potential be put into electric communication with another body, a flow of electricity takes place, the direction of flow depending on the state of the charge. Thus, if an electrified uninsulated conductor be charged with free positive electricity, this being at a higher potential than the earth, the direction of flow is from the body to the earth; if charged with free negative electricity, this is at a lower potential than the earth, and hence electricity will flow from the earth to the body.

We can consider a mass of water raised to some height above sea-level, and therefore possessing potential energy: if allowed to descend, the water may be made to do work in virtue of its energy, the amount depending upon the height and the mass. In the same manner, the higher the potential of an electrified body, the greater the amount of work it may be made to do in passing from the body to the earth. The negative potential would correspond to some level below sea-level, such as a well; so that if communication be made between the two, water flows in the reverse direction.

If the two bodies are at the same potential, no electricity

will pass from one to the other, as in the case of two bodies of the same temperature no flow of heat occurs. Thus the *potential* may be taken to mean the electric level above or below that of the earth, which is taken as zero. Potential above that of the earth is called positive (+), and below that of the earth negative (−).

Electric Density.

APPARATUS.—Electroscope; electrophorus; proof plane; conductors of various shapes.

EXPERIMENT 174.—Insulate a polished metallic sphere, A (Fig. 161); charge it by a few sparks from the electrophorus. If the proof plane be placed upon the surface of the sphere, it will receive a charge from it. Touch the knob of the electroscope, and note the divergence of the leaves; discharge both the proof plane and the electroscope; touch a different place on the surface

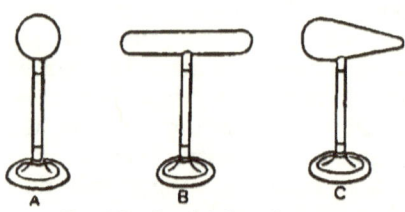

FIG. 161.—Insulated conductors.

of the sphere. Again charge, and note that the divergence of the leaves of the electroscope will be the same as before. If necessary, an electroscope with a graduated scale behind the leaves may be used. Repeat for several different parts of the surface, and it will be found that the electrification is the same at all points on the surface; in other words, the electrification on the sphere is uniformly distributed.

Next use a pear-shaped conductor, C, and it will be found that the electrification is greater at the ends than at the middle. Next use a flat disc, also any conductor which terminates in a blunt point; ascertain the difference between the amount in the case of a body terminating in a point, and in that of a similar body with rounded ends.

The term **electric density** is used to denote the *amount of electrification*—that is, *the number of units of electrification per square unit* (centimetre or inch) *of the surface.*

Condensers.—If an insulated conductor be separated by a non-conductor from another conductor, the arrangement will form a *condenser.*

The **Leyden jar** is a very convenient form of condenser. It consists of a wide-mouthed glass jar or cylinder (Fig. 162), coated

DIFFERENCE OF POTENTIAL—ELECTRIC DENSITY. 201

both inside and outside with tinfoil to within 2 inches of the top, then varnished inside and outside with shellac varnish.

A piece of brass wire having a knob at one end is fixed into a piece of wood, and the wood is placed at the bottom of the jar, and held in position by mixing some plaster of Paris with water and pouring the paste into the cylinder, holding the wire upright until the plaster sets. Inside the jar some crumpled tinfoil is placed, to make electrical connection between the wire and the inner coating.

That the apparatus may be used to store a relatively large quantity of electricity may be shown by the following experiment.

APPARATUS.—Leyden jar; electrophorus.

EXPERIMENT 175.—Place the jar on an insulated support, and charge it by passing a few sparks from the electrophorus.

FIG. 162.
Leyden jar.

The jar is discharged when conducting communication is made between the knob and outside. To discharge, what are called *discharging tongs* may be used. These may be simply a gutta-percha-covered copper wire about 2 feet long, having at each end a leaden bullet fused on, and bent as shown in Fig. 163.

To discharge the jar by means of the discharging tongs, place one leaden ball against the outside of the jar and bring the other near to the knob; the opposite charges in the jar will at once unite, and neutralize each other. A comparatively feeble spark is obtained.

FIG. 163.
Discharging tongs.

Repeat the experiment; lift the jar by the knob, and if one end of the discharging tongs be put into contact with the knob when the other end is put near to the outside coating, a strong spark is obtained.

Repeat, placing the jar on the table, or holding it in the hand whilst charging.

SUMMARY.

Potential.—Bodies are said to be of the same potential if, when placed in electrical communication, they neither give nor receive electricity from each other. When bodies are placed in electrical communication, they tend to become of the same potential. If when placed in communication positive electricity passes from the first to the second, the first is said to be at a **higher potential** than the second.

The potential of the earth is zero. A body charged with positive elec-

202 DIFFERENCE OF POTENTIAL—ELECTRIC DENSITY.

tricity is at a higher, and with negative electricity at a lower, potential than that of the earth.

Electric Current.—If two places are kept at different potentials, and are joined by a wire so that a continuous flow of positive electricity passes from one to the other, such a flow forms an *electric current*.

Surface density at a given point is the quantity of electricity per square cm. at that point.

A Leyden jar usually consists of a glass bottle lined inside and out to within a few inches of the top with tinfoil, forming two conductors separated by a non-conductor.

EXERCISES.

1. On touching the top of a charged Leyden jar standing on the floor or on a common table, you get an electric shock; but if either you or the jar stand on a dry cake of resin, you do not get a shock on touching the knob. Explain this.

2. Describe the construction of the Leyden jar, and give reasons, based (1) on experiment and (2) on theory, for believing that it can be used to store up a relatively large charge of electricity.

3. A large Leyden jar, the outer coating of which is earth-connected, is charged. If you wish to discharge it with the electric tongs, which coating should you touch first with the tongs, and why?

4. The extremity B of a wire A B is attached to the plate of a gold-leaf electroscope. By means of an insulating handle, the other end, A, is placed in contact first with the blunt and then with the more pointed end of a pear-shaped, insulated, and electrified conductor. Describe and explain the movements of the leaves of the electroscope.

CHAPTER XXIV.

VOLTAIC ELECTRICITY.

Simple Voltaic Cells—Electric Current.—If a plate or strip of pure zinc and a plate of ordinary commercial zinc be placed in dilute sulphuric acid, the impure or commercial zinc is attacked at various points along its surface, the zinc being eaten away by the acid at those points (and hydrogen set free). This so-called local action is caused by the presence of iron, etc., in the impure zinc. No such action occurs with the pure zinc, and by rubbing the surface of the impure zinc with mercury (amalgamated zinc) it is made to act like pure zinc.

To amalgamate the surface of a plate of ordinary commercial zinc, dip the zinc plate into dilute sulphuric acid—that is, water to which about an eighth part (by measure) of strong acid is added; when adding the acid, the water should be stirred with a glass rod. After the zinc has been acted upon by the acid for a minute or two, rub mercury over its surface by means of a piece of cloth or a piece of cork. The effect produced is to make the impure or commercial zinc, when placed in sulphuric acid, to act as pure zinc.

APPARATUS.—Glass beaker; dilute sulphuric acid; strips of copper, of pure zinc, and commercial zinc.

EXPERIMENT 176.—Place the strip of commercial zinc in the dilute acid, and note the bubbles of gas given off. Repeat the experiment, using strips of pure zinc and copper. No bubbles of gas are given off at the surface of either strip; but if the metals are connected either inside or outside, *bubbles of gas are given off at the copper strip.*

APPARATUS.—Strips or plates of amalgamated zinc; dilute acid; and beaker, as in last experiment.

EXPERIMENT 177.—Place a plate of amalgamated zinc and a copper plate in the dilute acid. It will be found that so long as the zinc and copper are not in contact, *no* bubbles of gas are given off from either metal; but if the plates are brought into contact (or connected by a wire), either inside or outside the vessel in which they are placed, a change occurs. Bubbles of gas are given off, *not from the zinc*, but *from the copper plate.*

A convenient method of attaching the two plates is by means of copper wire fastened to each, and the ends connected together.

APPARATUS.—As in last experiment.

EXPERIMENT 178.—After the two strips have been in contact for a few minutes, remove the strips, carefully wash and dry them, and obtain the weight of each. Replace and reconnect, and after being in contact for a short time (say 15 to 20 minutes), again remove, wash, and dry them, and weigh. It will be found that the *copper has not*, but the *zinc has, lost weight.*

Record your results as follows:—

Strips of copper and zinc in dilute acid.	At beginning.	At the end of 20 minutes' action.
Weight of zinc strip............	....	....
Weight of copper......	....	....

APPARATUS.—Zinc and copper plates; dilute acid; and beaker, as in last experiment.

EXPERIMENT 179.—(*a*) Underneath the copper wire connecting the zinc and copper plates place an ordinary compass needle (consisting of a light magnet suspended at its centre of gravity, and free to move in a horizontal plane). Note the deflection of the needle. (*b*) Wrap the copper wire a few times round a bar of soft iron, and note that iron filings are attracted. Also note, in (*a*), that the effect produced on the needle becomes less and less, until it finally ceases. *On cleaning and replacing the copper plate, the action is renewed.*

The peculiar properties of the wires are due to an electric current which *flows from the copper to the zinc plate.* This method of obtaining an electric current is due to an Italian philosopher named Volta. Vessels fitted in such a manner that an electric current is produced are called *voltaic* or *galvanic cells*. The points of attachment of the wires to the metal plates are called the *poles of the cells:* that of the copper is called the *positive pole*, the other

VOLTAIC ELECTRICITY.

the *negative pole*—the whole arrangement being called a *simple voltaic cell*.

By means of a very delicate electroscope, the copper wire attached to the copper plate is found to be *positively electrified*, and that which is attached to the zinc plate *negatively electrified*. The wire will attract iron filings, or deflect the magnetic needle. The end of the wire attached to the copper plate is at a higher potential than that which is attached to the zinc; hence the current flows through the wire from the copper to the zinc. The chemical action in the cell starts at the zinc or negative pole, and is transmitted from particle to particle of the liquid until the copper plate is reached. The copper plate (or its equivalent in other cells), where the chemical action inside the cell ceases and the external current along the wire begins, is the *positive pole*. The strength of the current depends upon the difference of potential. The end of the wire attached to the copper plate is at a higher potential than that attached to the zinc plate.

Voltaic Cells.

Polarization.—If, instead of the copper plate, carbon or platinum were used, the wire attached to the carbon or platinum would be positive, that attached to the other negative. (Salt water and iron could be used to replace the dilute sulphuric acid and the zinc respectively.)

It is found that the current in the voltaic cell just described rapidly diminishes, owing to the accumulation of hydrogen on the copper plate. The negative condition of the copper surface becomes neutralized by the film (positive) of hydrogen on it. This gradually lowers the intensity, and the current ceases. The negative metal is then said to be **polarized.** Polarization is prevented by—

1. **Mechanical method**—having a rough surface, so that the bubbles easily detach themselves from it; that is, roughening the surface of the copper by scratching with emery, etc.

2. **Chemical method**—by using some substance such as bichromate of potash or nitric acid, which are strongly oxidizing agents.

3. **Electro-chemical**—by the use of copper sulphate solution, which will attack the hydrogen and deposit the free copper thus formed on the copper plate.

Smee's Cell.—If platinum be deposited on a silver plate, the plate is said to be platinized, and the surface is very rough. By the substitution of such a plate for the copper plate in the simple voltaic cell, the bubbles of gas do not remain on the surface, but rise from the various points of roughness and escape at the surface.

In what is called a **Daniell's Cell**, the dilute sulphuric acid is placed in a porous pot, P (Fig. 164), which is enclosed in a copper,

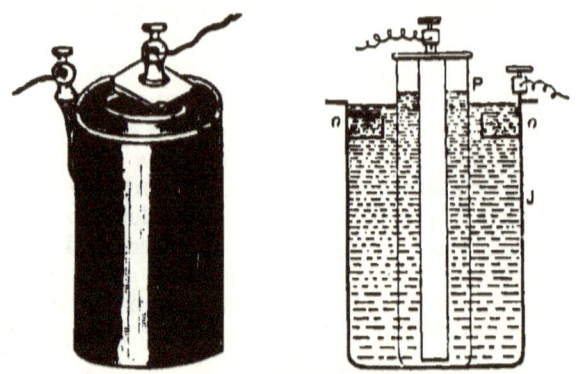

FIG. 164.—Daniell's cell.

glass, or glazed earthenware beaker or jar, J, of larger size—the space between the porous pot and the beaker being filled with a concentrated solution of copper sulphate, into which a strip of copper is inserted, forming the positive pole; the negative pole being a well-amalgamated rod or strip of zinc. When the terminals are connected, the dilute sulphuric acid acts upon the zinc, forming sulphate of zinc and liberating hydrogen. The reaction may be represented thus—

$$H_2SO_4 + Zn = ZnSO_4 + H_2.$$

The hydrogen attacks the copper sulphate, liberating the metallic copper and forming sulphuric acid. The reaction is—

$$CuSO_4 + H_2 = H_2SO_4 + Cu.$$

The liberated copper is deposited upon the copper plate, thus preventing polarization.

The copper sulphate would get weaker; hence some spare crystals of the substance are contained in a perforated cavity, as shown at *o* (Fig. 164).

APPARATUS.—Daniell's cell.

EXPERIMENT 180.—Connect the poles by a long, straight copper wire. Hold the wire over a magnetic needle in such a way that the current in the wire flows from north to south, and note the deflection of the needle. Next place the wire under the needle.

The needle will be deflected, the north pole of the needle being deflected towards the east; in the former case the deflection will be towards the west. If, without altering the position of the wire, the battery connections are altered so that the current flows from south to north, the results in both the above cases will be reversed.

The direction of deflection of the needle, in any case, can be found from **Ampère's Rule**—namely, *Imagine a man swimming in the wire in the direction of the current, with his face always turned towards the needle; then the* **north pole** *is in every case deflected towards his* **left hand.**

The **Grove's Cell** has a higher voltage and less internal resistance than the Daniell's cell, and consists of an outer cell of glazed earthenware or of ebonite, which contains the dilute sulphuric acid and the amalgamated zinc plate. The inner porous cell contains the strongest nitric acid, and the negative pole is furnished by a very thin plate of platinum, placed in the nitric acid.

Leclanché's Cell is very largely used for all purposes requiring an intermittent current, such as electric bells, telephones, etc. The dilute acid is replaced by a solution of sal ammoniac. The cell consists of a rod or slab of zinc, Z, carbon, C, and powdered black oxide of manganese (Fig. 165). The carbon and manganese are placed in the porous vessel, P; a certain amount of sal ammoniac is placed in the outer vessel, and water poured on it.

When used to give a continuous current for several minutes, the power rapidly falls off, owing to the accumulation of hydrogen; but when left to itself for a short time, it rapidly recovers.

FIG. 165.
Leclanché's cell.

Its advantages are—it does not require renewal for months or years, and does not contain corrosive acids.

The action of the liquid on the zinc rod causes hydrogen to form, but this combines with some of the oxygen from the oxide of manganese, and water is formed; thus the manganese compound is gradually reduced.

EXPERIMENT 181.—Connect two copper plates of the same weight (using platinum wire terminals) with the positive and negative poles respectively of a Bunsen cell, and immerse side by side in a solution of copper sulphate containing a small quantity of sulphuric acid. After being in action for some time, remove the plates, dry, and reweigh them. The weights will no longer be alike. The copper plate connected to the negative pole is found to be heavier than before, on account of the copper deposited (the difference in weight gives the amount); the one connected to the positive pole weighs less.

Replace the two plates, but reverse their connections with the poles of the battery, and allow them to remain for the same time as before. Again remove carefully, dry, and re-weigh them, and the original equality in the weights will be found. Write out fully your explanation of this.

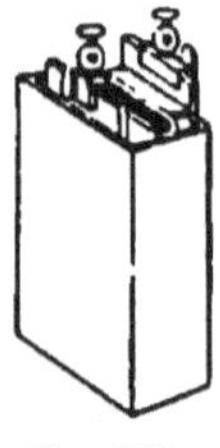

Bunsen's Cell.—In this cell, which is very often used, a slab or rod of carbon replaces the expensive platinum of the Grove's cell.

The stoneware jar (Fig. 166) is filled to within about an inch of the top with dilute sulphuric acid, and the porous inner pot with strong nitric acid.

FIG. 166.
Bunsen's cell.

A convenient form of Bunsen cell is shown in Fig. 167. The sheet of zinc, Z, is bent into a cylindrical form, and placed within a stoneware jar, J, as shown. A porous pot, p, is placed inside the cylindrical piece of zinc, and inside this a slab or square rod of compressed carbon, C. Binding screws to connect the wires are clamped to the zinc and to the carbon.

FIG. 167.

EXPERIMENT 182.—Connect two copper plates as in the last experiment, and place them in the solution. After being in action for a short time, the copper plate joined to the zinc of the cell will be found to be covered with bright copper. The plate of the other pole will be partially dissolved.

If the two plates be dipped into dilute sulphuric acid, it will be seen that gas is given off at both poles.

If the gases were collected separately, they would be found to be oxygen and hydrogen. Oxygen is given off at the pole which is joined to the carbon of the cell, and that given off by the other pole is hydrogen. As the oxygen given off acts upon the copper, it is better to use platinum terminals.

The volume of the oxygen given off at the positive pole is half that of the hydrogen given off at the negative pole.

Any substance which is decomposed by the passage of an electric current is called an **electrolyte**; the action which takes

VOLTAIC ELECTRICITY.

place is called **electrolysis**; the two poles suspended in the electrolyte are called **electrodes**. The pole at which the current enters is called the **anode**, and that at which it leaves the **kathode**.

Heating Effect of Current.

APPARATUS. — Two Bunsen cells; a few inches of fine platinum wire.

EXPERIMENT 183.—Twist the two ends of the platinum wire, and connect them to the two pieces of copper wire. Connect to the two cells so that a current is made to pass through the wire. It will be found that the wire gets hot. If immersed in a known quantity of water, and the temperature of the water, before and after the current is passed, be noted by means of a thermometer, then the heat given to the wire in that portion of its circuit can easily be estimated.

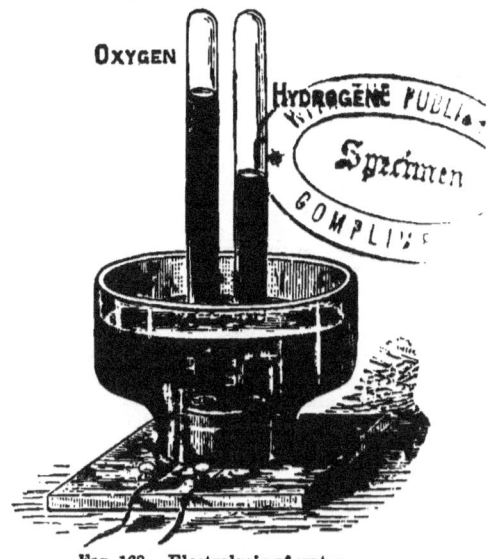

FIG. 168.—Electrolysis of water.

Proceed to shorten the wire, and note that it gets hotter and hotter, until it becomes white-hot, giving out a bright light. If the shortening be continued, it is possible to fuse the wire.

Hence *an electric current passing through a conductor heats it.*

SUMMARY.

Simple Voltaic Cell.—Commercial zinc is, but pure or amalgamated zinc is not, acted upon by dilute sulphuric acid. When strips of zinc and copper are placed in dilute sulphuric acid and joined by a wire, positive electricity flows from the zinc to the copper, and from the copper through the wire to the zinc. The zinc is eaten away, and hydrogen bubbles appear on the copper plate.

Electric Current.—The end of the wire connected to the *copper strip* is at a *higher potential* than the end attached to the *zinc;* hence an *electric current* passes along the wire, due to this difference of potential.

Polarization.—The wire connecting the zinc and copper strips is found to attract iron filings or to deflect a magnetic needle. This action becomes weaker, and after a time ceases, the negative condition of the copper surface being neutralized by a film of hydrogen on it. The negative metal is then said to be polarized.

Prevention of Polarization.—(1) By mechanical, (2) by chemical, (3) by electro-chemical methods.

Smee's Cell.—A platinized silver plate is substituted for the copper plate in the simple voltaic cell.

Daniell's Cell.—The hydrogen acts upon copper sulphate, liberating the copper, which collects on the copper plate, but does not interfere with the action.

Grove's Cell.—A strip of platinum foil replaces the copper of the simple cell, and nitric acid is used: the "polarizing" hydrogen is oxidized to water.

Bunsen's Cell.—A slab or rod of carbon replaces the expensive platinum of the Grove's cell.

Leclanché's Cell.—The dilute acid is replaced by a solution of sal ammoniac. This acts on the zinc rod, liberating hydrogen; but the hydrogen is oxidized by the black oxide of manganese. This cell cannot be used where a continuous current is required, but it is very useful for electric bells, etc.

The passage of the electric current through a conductor heats it. Certain liquids, as turpentine, will not conduct the current.

Electrolysis.—When an electric current is passed through water, the water is decomposed into its chemical elements, oxygen and hydrogen—the former appearing at the positive, and the latter at the negative pole, the volume of the hydrogen being twice that of the oxygen.

Electrolyte is the name given to a liquid which conducts the current and is decomposed by it. The ends of the wires which come from the poles of the battery are called **electrodes**; that connected to the positive pole is called the anode, and that connected to the negative pole the **kathode**.

Exercises.

1. Give a drawing of a galvanic cell of copper, zinc, and dilute sulphuric acid, showing in what direction the positive current passes through a wire connecting two metals, and also through the dilute sulphuric acid.

2. What are the materials used in the construction of a Daniell's cell, and what chemical changes occur in the cell when in action?

3. Explain why, when a sufficiently strong current passes through an incandescent lamp, the lamp becomes hot, while the wires which lead the current to it are comparatively cold.

4. Plates of platinum and copper are dipped into a solution of copper sulphate. What effects are produced upon them if a current is passed through the liquid from the copper to the platinum?

5. A horse-shoe-shaped piece of copper is hung from the arm of a balance, so that the two legs dip into two different vessels containing a solution of copper sulphate, and weights are placed in the other pan till the balance is in equilibrium. An electric current is passed for some time through the vessels and the horse-shoe, the electrodes being copper plates dipping into the liquid in the vessels, but not touching the horse-shoe. When the current is stopped, will the balance still be in equilibrium? Give reasons for your answer.

CHAPTER XXV.

ELECTRO-MAGNETISM.

APPARATUS.—A strip of iron or steel having several turns of cotton-covered copper wire wrapped round it so that the ends are bare.

Electro-Magnets.

EXPERIMENT 184.—Suspend the coil and the strip by silk string (Fig. 169). Pass a current from the Bunsen cell through the coil: the strip becomes magnetized so long as the current is flowing. This is shown by its power of attracting iron filings or small pieces of metal, such as tacks, iron nails, etc., when placed near the ends of the strip.

Reverse the current by interchanging the two wires connected to the poles of the cell; ascertain the alteration which occurs in the polarity of the coil and the enclosed strip.

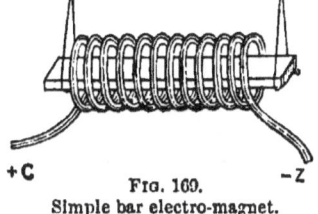

FIG. 169.
Simple bar electro-magnet.

The bar of soft iron, which thus becomes a temporary magnet, is called an **electro-magnet**.

Solenoid.—If the length of the coil of wire be many times its diameter, it is called a *solenoid*. To ascertain polarity of an electro-magnet, Ampère's rule may be applied. Thus, *a person swimming in the wire in the direction of the current, and facing the needle, will have the north pole on his left hand;* or, the polarity of the coil is so related to the direction of the current, that looking towards the south end of the helix, the current circulates in the same way as the hands of a watch move.

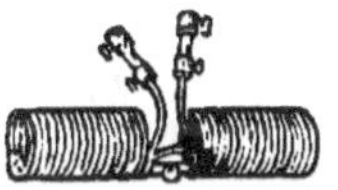

FIG. 170.—Solenoid.

The polarity of the coil being the same as that of the iron

inside, the arrangement of the solenoid and its core is called an **electro-magnet**.

A solenoid through which an electric current is made to pass behaves exactly like a magnet if freely suspended: it sets in the magnetic meridian, and is repelled and attracted by another solenoid or permanent magnet.

APPARATUS.—Concentric mercury cups; covered copper wire.

EXPERIMENT 185.—Take a piece of iron wire and wrap round it insulated copper wire, as shown in Fig. 171. Bend the wire so that the ends just pass under the surface of mercury contained in concentric grooves in the wood block B.

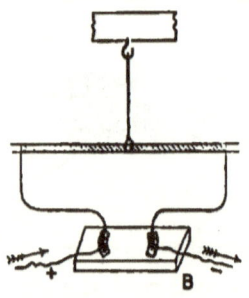

FIG. 171.—Solenoid.

Suspend the coil and the enclosed iron wire by untwisted silk. Put one terminal of a voltaic battery (several cells needed) in each groove containing mercury: it will be found that the coil and rod will swing into a position nearly north and south—showing that the rod has become a magnet.

Interchange the two wires connected to the poles of the battery so as to reverse the current; ascertain what effect this produces on the polarity of the iron core.

Current.—It has been shown in a simple voltaic cell (Experiment 169) that the action is from the zinc, through the liquid to the copper. This action is called *current;* the cause of the current, due to chemical action in the cell called electro-motive force (written E.M.F.), is *proportional to the difference of potentials between the two poles.* Difference of potential also depends on the materials used. Thus, when sulphuric acid of the same strength and temperature is used, and the cell consists of zinc, copper, and acid, a certain value (always the same) is obtained; but the E.M.F. of the cell is altered if iron replaces the zinc, or platinum the copper, etc.

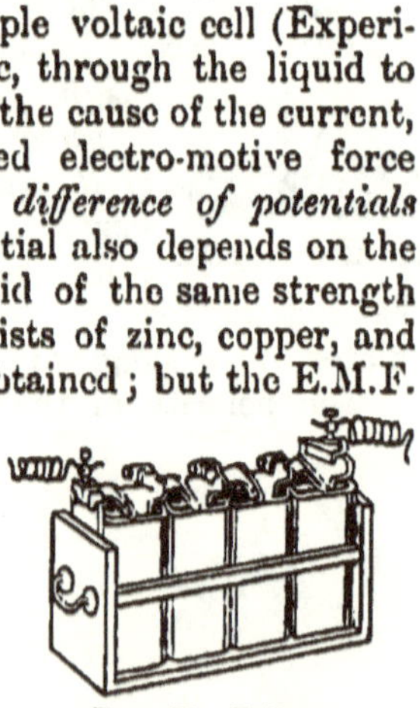

Battery.—To produce a stronger current than that furnished by one cell, a number of cells may be joined together; such a combination (Fig. 172) is called a *battery.*

FIG. 172.—Battery.

There are two methods of joining the cells together. Thus, if all the copper (or +) poles and all the zinc (or −) poles be joined together, the E.M.F. is not altered; the plates being joined together in this manner form one conductor, the difference of potential remains the same as that of a single cell, and *the cells are said to be* **joined in parallel.**

But by joining the + pole of one cell to the − pole of the next, and so on, the difference of potential of any cell is added to the one next to it. Thus the difference of potential and the E.M.F. are increased: *this is called* **joining the cells in series.**

The unit of E.M.F. is called the *volt*, the **volt being** approximately $\frac{11}{10}$ the **E.M.F.** of a single Daniell cell.

Galvanometer.

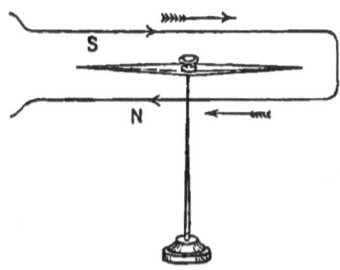

Fig. 173.—Wire carried above and below a magnetic needle.

By *Ampère's rule* on p. 211, the direction of an electric current flowing through a wire can easily be determined.

Thus, if a freely-suspended magnetic needle be used, above which, and parallel to it, a wire is placed through which a current is passing, the needle will be deflected from the magnetic meridian in which it rested. If, for instance, the north pole was found to be deflected towards the west, then we know from Experiment 180 that the current in the wire is from south to north; also the deflection of the needle would indicate roughly the strength of the current. Such an arrangement constitutes a **galvanometer** in its simplest form.

If the wire be bent into the form of a rectangle, as shown in Fig. 173, both portions of the wire by *Ampère's rule* tend to turn the north pole towards the west; hence the power of the current to turn the needle is doubled.

When fine wire covered with some insulating material—gutta-percha, cotton, or silk—is carried both above and below the needle a great number of times, forming a coil of many turns, the power to turn the needle is correspondingly increased. The needle lies inside the coil, in which it is free to turn, the adjustment being made by three levelling screws, as shown in Fig. 174. A light pointer ab

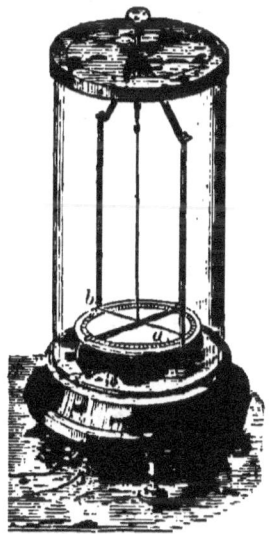

Fig. 174.—Galvanometer.

attached to the needle moves over a graduated dial as shown, and indicates the deflection of the needle. A glass cover protects the whole apparatus, and carries at the top a small adjusting screw to allow the pointer to swing freely, or, when not in use, to rest on the dial.

In the **Astatic Galvanometer**, two needles, of the same length and power, are fastened together in the same vertical plane, one inside, the other outside the coil (Fig. 175), having the north end of one needle opposite the south end of the other. We have already found that when a magnetic needle is freely suspended, it tends to set in the direction of magnetic north and south, owing to the earth's force (as a large magnet). And before a single needle can be deflected, this force has to be overcome. In the astatic pair of needles this force is neutralized (or nearly so); thus the force of the current swings the needle much more readily than before, and hence a weak current can produce a comparatively large deflection.

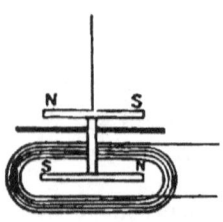

FIG. 175.
Astatic galvanometer.

Resistance.—We have found that some substances readily transfer electrification from electrified bodies, and offer very little resistance to the flow of electricity. These are called *conductors*, to distinguish them from other bodies, which offer such great resistance as practically to allow no electricity to pass through them, or only transfer it in a slight degree: such substances are called non-conductors or *insulators*. It should be noted, however, that every so-called *conductor offers some resistance*, and *every insulator allows some electricity to pass through it.* Whether electricity will or will not pass through a substance depends upon the E.M.F. and the resistance which the substance offers to its passage.

As already indicated, the resistance which the substance offers to the flow of electricity causes the development of heat; and knowing the work equivalent of heat as estimated by Joule, the power of an electric current can be estimated.

When the resistance is very great, such as that offered by the fine carbon filament inside an electric lamp, the amount of heat developed is so great that the filament becomes white-hot or incandescent.

If a rod of carbon, thick at one end and tapering towards the other end, be connected by copper wires to the poles of a powerful battery, the thicker portions may remain unaltered in appearance, but the thinner portions will be heated to redness. The current at any given instant is the same at each section. The quantity of heat developed at any section is proportional to the resistance at that section, but the resistance of unit length of the thin part is greater than the corresponding resistance of an equal length of the thick part; and as the heat conductivity of carbon is small, the heat is not readily distributed.

APPARATUS.—Galvanometer; Daniell or other cell; several pieces of wire of different materials and thickness.

EXPERIMENT 186.—Join, by means of a short length of copper wire, one pole of the cell to the galvanometer, and connect the other pole by means of about 6 or 8 feet of No. 30 copper wire; note the deflection produced. Shorten the long length of wire to half or quarter of its former amount, and again note the increase in the deflection, due to shortening the wire. Replace the wire by another of similar material but larger in diameter; you will find the deflection to be greater than with a corresponding length of thin wire.

Repeat the experiment, using No. 30 iron wire instead of the copper; and again repeat, using German silver wire. You will find that the copper wire transmits the stronger current—this being indicated by the larger deflection of the needle.

Enter all the details of the experiment in your note-book, and ascertain which metal offers the greatest resistance.

The resistance of a wire depends on—(*a*) length; (*b*) sectional area; (*c*) temperature; (*d*) material.

∴ Round wires of diameter, d, the resistance, r, is proportional to $\dfrac{1}{d^2}$.

The practical unit of resistance is the ohm, named after Dr. Ohm, who pointed out that the *strength of a current varies directly as the total electro-motive force, and inversely as the total resistance of the circuit*. Example:—If the electro-motive force is 60 volts, and the resistance of the wire 5 ohms, the current is $\frac{60}{5} = 12$ ampères. Similarly, if the electro-motive force be 1 volt and the resistance 1 ohm, then the current is 1 ampère; and the *ampère* could be defined as *the current due to a pressure of 1 volt, and through a resistance of 1 ohm*.

Thus, if E denote the electro-motive force in volts,
    C the current in ampères,
    R the resistance of the circuit in ohms,
then $C = \dfrac{E}{R}$.

If the resistance of silver wire be ·5 ohm, the resistances of copper and iron are ·505 and 3·12 respectively.

It should be noticed that the resistance = sum of resistances of its various parts *in series* ("in series" simply means that the current has to pass through all the parts consecutively).

The expression

$$\text{Strength of current} = \frac{\text{Electro-motive force}}{\text{Total resistance of circuit}}$$
$$= \frac{E}{R}$$

is known as *Ohm's Law*.

Thus, if the circuit were composed of cell E.M.F. = E, and resistance Rb; galvanometer resistance Rg, and the connecting wires' resistance Rw;

$$\text{then } C = \frac{E}{Rb + Rg + Rw}.$$

From which, when E and the resistances are known, the current can be ascertained.

Dr. Joule found that the *E.M.F. of a Daniell cell was 1·039 volts*, or approximately 1 volt.

The *E.M.F. of a Bunsen cell is approximately 2 volts.*

In the Daniell and the Bunsen cell, the energy given out is due to the combustion of zinc by the sulphuric acid; and when unit quantity of electricity passes through the circuit, ·00033 grams of zinc is consumed. But the work done in the Bunsen is twice as great as in the Daniell cell.

Quantity and Current.—*Current* is the rate at which electricity flows through wires, and is measured in **ampères**.

If the E.M.F. be 1 volt, and the resistance 1 ohm, the current is 1 ampère.

The **unit of quantity** is called a **coulomb**. It should be noted that just as the number of gallons of water in a tank has nothing to do with the head of water (due to height of tank), or with the velocity of the water in the pipes leading from the tank, so *coulombs* have nothing to do with the number of volts, or with the current in ampères. The quantity of water flowing in a current may be measured as so many cubic inches per second. So the quantity of electricity in a current is the quantity per second flowing along the conductor, and is measured in coulombs. If the passage of electricity through a wire be at *the rate of* **one coulomb per second**, it is called the rate of **one ampère**, or a current of one ampère.

ELECTRO-MAGNETISM.

If a steady current is flowing, the quantity that has passed in a given number of seconds is found by multiplying the current (in ampères) by the number of seconds.

$\therefore Q = Ct$, where Q = quantity.
C = current in ampères.
t = time in seconds.

In the C.G.S. system, the unit of work is one **erg**, and is the work done by a force of one dyne through a distance of one centimetre; also 10^7 ergs per second is the electrical unit of power, and is called a **watt**.

It is easy to ascertain the number of watts equivalent to one horse-power. Thus:—

1 pound = 453·6 grams.
1 foot = 30·48 centimetres.
$\therefore$ 1 H.P. = 550 foot-pounds = 550 × 453·6 × 30·48.
$\therefore$ 1 H.P. = $\dfrac{550 \times 453 \cdot 6 \times 30 \cdot 48}{10^7}$ = 746 watts.

This relation is very important when required to *measure H.P. electrically*. *The current in ampères is multiplied by the E.M.F. in volts, and divided by 746.*

Wheatstone's Bridge.—Resistance can be measured in a simple manner by means of a Wheatstone's bridge. If the current in a wire, W, flowing in the direction indicated by the arrow—that is, from place of higher to that of lower potential—is divided at O, and to reach the wire at C is made to pass along the two wires or conductors O B C and O A C, the currents flowing along these wires will be inversely as the resistances of the conductors, the fall of potential being as much for one conductor as for the other. There is corresponding to some point, B, on one conductor, some point, A, on the other having an equal potential to that of B; hence if A and B were joined by a conductor, as the potential of each is the same, no current would flow through it, and any current sent through W would not affect the most sensitive galvanometer attached to the wire.

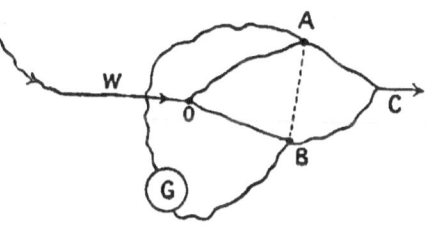

Fig. 176.

There are two forms of Wheatstone's bridge, known as the *coil* and the *wire* bridge respectively; O A, O B, and A C are three sets of resistance coils, and BC (Fig. 177) is the wire the resistance of which is to be measured.

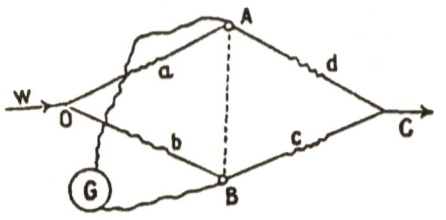

Fig. 177.—Wheatstone's bridge.

The two lengths OA and OB are called the arms of the bridge, and are either equal in length or in some definite known proportion to each other.

The resistance AC is adjusted until no deflection is observed on the galvanometer, G. Denoting the resistances of a, b, c, and d by Ra, Rb, Rc, and Rd respectively,

Then Ra : Rb : : Rd : Rc.

$$\therefore Rd = \frac{Ra \times Rc}{Rb}.$$

When a and b are equal, then Rd = Rc.

APPARATUS.—Wheatstone's bridge; galvanometer; wires of various materials and of different sectional areas; Daniell cell.

EXPERIMENT 187.—Find as accurately as you can the resistances of equal lengths of the wires supplied, by inserting them in the gaps provided; also find, by using the same materials and length, but different sectional area, how the resistance varies with the area (and diameter).

Find, by using wires of the same thickness, the lengths of each of the following to give the same resistance: German silver, copper, iron, and so on.

Enter all the details in your note-book.

SUMMARY.

Electro-Magnet.—If a bar of soft iron be placed inside a coil of wire and an electric current sent through the coil, the iron becomes magnetized, but ceases to be a magnet when the current ceases. If steel be used instead of the soft iron, the steel becomes a permanent magnet.

Ampère's Rule.—Imagine a man swimming in the wire in the direction of the current, having his face always turned towards the needle: the north end of the needle will always be deflected towards his left hand.

Solenoid consists of a movable wire helix, through which an electric current is made to pass. The solenoid behaves like a magnet.

Volt is the unit of E.M.F.

A Galvanometer is an instrument for measuring the intensity of an electric current by the deflection of a magnetic needle.

Astatic Galvanometer consists of two equal needles, one fastened outside, the other inside a coil of wire—the north end of one needle being opposite the south end of the other; in this form a weak current is able to produce a comparatively large deflection of the needle.

Resistance of a wire depends upon its (1) *length*, (2) *sectional area*, (3) *temperature*, (4) *material*.

Ohm's Law.—Strength of a current is equal to the electro-motive force ÷ total resistance of circuit.

Ampère.—The rate at which electricity flows through wires is measured in *ampères*. If the E.M.F. be 1 volt and resistance 1 ohm, the current is 1 ampère.

1 *horse-power* is 33,000 foot-pounds of work per minute.

1 *Watt* is 10^7 *ergs* per second.

1 H.-P. = 746 watts.

The product of *current in ampères* by *E.M.F. in volts* ÷ 746 is used to measure H.-P. electrically.

Wheatstone's Bridge, an instrument used for measuring resistance of a wire.

Exercises.

1. Describe the construction of the most sensitive galvanometer with which you are acquainted.

2. Why is an astatic galvanometer better adapted for the measurement of weak currents than a galvanometer with a single needle?

3. How would the substitution of dilute sulphuric acid for strong nitric acid in a Grove's cell affect (1) the value, (2) the constancy of its E.M.F. when working?

4. When a current passes along the wire joining the terminals of a battery, does it also pass through the battery? Give reasons for your answer.

5. A strip of platinum and a strip of zinc dip into a vessel of acidulated water. How would you show that two copper wires, fastened one to the zinc and the other to the platinum, are in different electrical states?

ANSWERS TO EXERCISES.

CHAPTER II.—(Page 29.)

1. 88 inches, taking $\pi = \frac{22}{7}$.
   87·9648, taking $\pi = 3·1416$.
   The first answer is correct to within $\frac{1}{70}$ inch.
2. 10 inches.
3. 20·11.
   In many cases it is not necessary to retain more than two decimal places in the result.
4. 9·5 cm.
5. 2309; 7752.

CHAPTER III.—(Page 36.)

1. 836·097.
2. 17·89 ft.

CHAPTER IV.—(Page 54.)

1. 2·21 lbs.
2. 2,760 kgm.
4. 17 lbs. 3 oz.
5. 135·9 lbs.
6. 312·5.

(Page 58.)

1. ·84.
2. ·786.
3. 7½ oz.
4. 10·63.
5. ·79.

(Page 61.)

2. ·27.
4. ·798 mm.
7. 69·4 lbs.

CHAPTER VI.—(Page 88.)

1. 55 ft. per sec.; 66,000.
2. (a) 40 ft. per sec.; 1⅝.
3. Poundal; 1⅞.
4. 11 ft. per sec.
5. 48.
6. 60; 5 : 32.
7. 5 ft. per sec.
8. 24,000 yds. a min. in one min.
9. ⅜ ft. per sec.
10. 10. 17·3; 20.
11. 1·85 v where v is vel. along A B.
13. 27; 33.
14. (a) 11; (b) 8½ ft. per sec.; 100°.
15. 27·8; 68° with A C.
16. 112° 37'; 90°.
17. 25 lbs.
18. 70·7; 25·9.
19. 20 ft. per sec. in a direction parallel to B C.
20. 22·4.
21. 21·8; 13·2.
22. (a) $\frac{22}{7}$, (b) 25·3, (c) 7·0 sec.
23. 7·84 seconds.
24. 172 ft. per sec.
25. 29·3 ft. per sec. per sec.; 0·945.
26. 64 ft. below point of projection.
27. 28·8 ft. per sec.
28. 12; 8.
29. 9 and 21; 10 lbs.
30. 1,190 lbs.; 1,140 lbs.
31. 13·9.
32. 10 ft.
33. 15 : 2.
34. 60, 20, 40, 60.
35. 7·407 ft. per sec.; 3,841,600 to 939,250,000 poundals.
36. Let 0 be point of inter., and A the angular point, G the centre of gravity; then $OG = \frac{1}{3} OA$.
37. $\frac{5}{12}$.
41. 2⅔ ft.; ½ lb.
42. At 0 point of intersection of diagonals; at middle point of 0 A.
43. 21 inches and 15 inches.

ANSWERS TO EXERCISES.   221

CHAPTER VII.—(Page 107.)

1. 56.
2. 30,000 foot-poundals; 312·5 ft.
3. 20,160 foot-pounds; 3,360 foot-poundals.
4. 500 foot-poundals; 16,000 foot-poundals; $v = 88$ ft. per sec.
5. 12,000 foot-poundals.
6. 28·28 ft. per sec.
7. 35·16.
8. $66\frac{2}{3}$.
9. $2\frac{1}{2}$ ft. per sec.; 147 ft.
10. 2,139 foot-pounds per min.; ·06 H.-P.
11. 0·67 H.-P.
12. 1,400 foot-pounds.
13. (a) 20,000; (b) 625.
14. 320.
15. 1,650; 20.
16. 108.
17. 1,600; 1,280.
18. 2,250; 30.
19. 26,400.
20. 2,240 foot-poundals; 44·8.
21. 1,080.
22. 192; 5 ft. per sec.
23. 16 hours.
24. 576.
25. 9,600; $2\frac{1}{2}$ ft. per sec.
26. 11 ft. per sec.

CHAPTER VIII.—(Page 115.)

1. 39·139 inches.
2. $9\frac{65}{121}$ ft.
3. 32.
4. 4·67 ft.
5. Time of one beat, 1·014 secs. Number of beats in twenty-four hours, 8,761.
6. 36·47 inches.
7. $f = \frac{1}{8}g$; $t = 1·22$ secs.
8. $f = 8$; $\dfrac{P}{Q} = \dfrac{5}{3}$.

CHAPTER IX.—(Page 123.)

1. $\frac{1120}{256} = 4·375$ ft., and $\frac{1132}{256} = 4·422$ ft.

CHAPTER X.—(Page 129.)

3. $35\frac{1}{5}$ ft. per sec.; rate of vibration, 496.
7. Wheel makes 4 revs. per sec.; ∴ frequency = $4 \times 64 = 256$.
8. 1,860.

Wave length $= \dfrac{1126·4}{256} = 4·4$ ft.

CHAPTER XI.—(Page 135.)

2. $\frac{18}{25}$.
3. $\frac{18}{81}$ I.

CHAPTER XIV.—(Page 156.)

1. ·0306 yds.
2. 0·043 inches.
3. 1·0057 cubic decimetres.
4. 0·0354 c.c.
5. 506·9 ft.
6. ·000016.

CHAPTER XV.—(Page 161.)

3. 356·26 lbs.
4. 1073·2 to 1917·73.

CHAPTER XVI.—(Page 167.)

1. 190°, 185°, 160°, 22°, −5°, −10°.
2. 500°, 193·6°, 160°, 135·2°, 249·8°, 30·2°, 19·4°.
3. $11\frac{1}{9}$° C.; $16\frac{2}{3}$° C.
4. $-6\frac{2}{3}$°.

CHAPTER XVII.—(Page 175.)

1. 90·9°.
2. 0·1007.
3. $\frac{1}{18}$.
4. 173·7.
5. $9\frac{2}{11}$° C.
7. 966·1.

TABLE OF USEFUL CONSTANTS.

1 centimetre = 0·394 inch.
1 metre = 3·28 feet.
1 inch = 2·54 centimetres.
1 foot = 30·47 centimetres.
1 kilogram = 2·2046 pounds.
1 gallon = ·16 cubic feet = 10 pounds of water at 62° F.
Value of g = 32·18 feet per second per second.
1 gram = g = 981 dynes.
1 atmosphere = 14·7 pounds per square inch = 760 millimetres of mercury.
π = 3·1416 or $\frac{22}{7}$.
π^2 = 9·87.
$\frac{1}{\pi}$ = 0·318.

Specific heat of copper = 0·095.
Specific heat of glass = 0·19.
Specific heat of mercury = 0·033.
Coefficient of cubical expansion of glass per 1° C. = 0·000026.
Coefficient of cubical expansion of mercury per 1° C. = 0·00018.
E.M.F. of a Daniell's cell, about $1\frac{1}{2}$ volts.

1 ohm is nearly equal to the resistance of 10 feet of pure copper wire 0·01 in diameter.

INDEX.

ABSOLUTE expansion of a liquid, 158.
Acceleration, 79, 80.
Actinic energy, 130.
Air, properties of, 71.
—— thermometer, 158.
Alcohol thermometer, 162.
Amalgamed silk, 192.
Ampère's rule, 207.
Amplitude, 110, 120.
Angle, measurement of, 33.
—— of deviation, 138.
—— of friction, 106.
—— of refraction, 138.
Apparent expansion of a liquid, 158.
Archimedes, principle of, 58.
Area, measurement of, 35.
—— unit of, 11, 12.
—— by weighing, 44.
—— of a circle, 38.
—— of a parallelogram, 37.
—— of a rectangle, 35.
—— of a square, 35.
—— of a triangle, 35.
—— of an ellipse, 42.
—— of an indicator diagram, 43.
—— of curved surface of cone, 41.
—— of curved surface of cylinder, 41.
—— of irregular figure, 40.
Astatic galvanometer, 214.
Attwood's machine, 113.
Average velocity, 80.

BALANCE, 49.
Barometer, 72.
—— Fortin's, 73.
—— Siphon, 74.
Battery, 212.
Boiling-point depends upon the pressure, 166.
Boyle's law, 74, 159.
—— law tube, 74.

British and metric measures compared, 14
Bunsen's cell, 208.
—— photometer, 133.
Burette, 45.

CALIPERS, 24.
Capacity for heat, 169.
Centre of gravity, 86.
Charles's or Dalton's law, 159.
Circle, area of, 38.
—— to measure circumference of, 28.
Cohesion, 16.
Compasses, 23.
Components of a force, 81.
Compound pendulum, 111.
Compressibility, 16.
Concave mirror, 148.
Condensers, 200.
Conduction of heat, 170.
Cone area of curved surface, 41.
—— development of, 32.
—— volume of, 47.
Conservation of energy, 93.
Convection, 170.
Coulomb, 216.
Couple, 84.
Critical angle, 140.
Cube, volume of, 47.
Cubical expansion, 155.
Curved line, measurement of, 32.
Cylinder area of curved surface, 41.
—— volume of, 46.

DANIELL's cell, 206.
Davy safety lamp, 178.
Declination, 187.
Density, 53.
—— of a liquid, 55.
—— electric, 200.
Development of cone, 32.

224 INDEX.

Diagonal scale, 23.
Difference of potential, 109.
Differential thermometer, 160.
Dip, 187.
Discharging tongs, 201.
Disc siren, 126.
Dividers or compasses, 23.
Divisibility, 16.

EFFICIENCY, 95.
Electric density, 200.
—— induction, 196.
Electrification, positive and negative, 192.
—— by contact, 193.
—— by friction, 191.
Electro-magnets, 211.
Electrophorus, 197.
Electroscope, Faraday's ice-pail, 198.
—— gold-leaf, 194.
Electrolysis, 209.
Electrolyte, 208.
Ellipse, area of, 42.
Energy, 92.
—— actinic, 130.
—— conservation of, 93.
—— potential, 92.
—— radiant, 130.
Equal expansion of gases, 159.
Estimation of force of gravity, 78.
Ether, 130.
—— waves, 130.
Expansion, cubical, 155.
—— linear, 152.
—— superficial, 155.
—— of gases, 158.
—— of liquids, 157.
—— of solids, 155.
Extension, 15.

FARADAY's ice-pail electroscope, 198.
Fixed points of a thermometer, 164.
Focal length of lens, 145.
Focus of lens, 145.
Force, components of, 81.
—— measurement of, 77.
—— measurement of, by tension, 77.
—— moment of, 83.
—— representation of, 77.
—— unit of, 79.
—— of gravity, estimation of, 78.
—— resultant, 81.
Forces, triangle of, 82.
Fortin's barometer, 73.
Frequency, 120.
Friction, 104.
—— angle of, 106.

GALVANOMETER, 213.
—— astatic, 214.
Gases, expansion of, 158.
—— equal expansion of, 159.
Gauge, 25, 27.
Glass, refractive index of, 141.
—— specific heat of, 170.
—— prism, refraction through, 144.
Gold-leaf electroscope, 194.
Graduating a thermometer, 165.
Gravity, centre of, 86.
Grease-spot photometer, 133.
Grove's cell, 207.

HARE's apparatus, 65.
Head of a liquid, 67.
Heat, 152.
—— capacity for, 169.
—— conduction of, 170.
—— latent, 172.
—— specific, 168.
Heating effect of current, 200.
Hope's experiment, 160.
Horse-power, 94.
Hydrometers, 68.
—— Nicholson's, 69.
Hydrostatics, 52.

IMAGES, 140.
—— multiple, 143.
—— size of, 149.
Impenetrability, 16.
Inclined mirrors, 143.
—— plane, 95.
Index of refraction, 139.
Indicator diagram, area of, 43.
Induction, 183.
—— electric, 196.
Inertia, 16.
Instruments used for measurement of length, 19.
Irregular figure, area of, 40.

JOLLY's balance, 62.

LABORATORY work, vii.
Latent heat, 172.
—— of steam, 173.
—— of water, 172.
Laws of inverse squares, 131.
—— of reflection, 136.
Lead, specific heat of, 169.
Leclanché's cell, 207.
Length, measurement of, 18, 21, 28.
—— unit of, 10, 12, 17.
—— instruments used for measuring, 19.
Lenses, 144.
—— focal length, 145.
—— principal focus of, 145.

INDEX. 225

Lever, 85, 97.
Leyden jar, 200.
Light travels in straight lines, 130.
Linear expansion, 152.
Lines of force, 182.
Liquid, absolute expansion of, 158.
—— apparent expansion of, 158.
—— density of, 55.
—— expansion of, 157.
—— head of, 67.
—— volume of, 49.
Litre, 13.
Longitudinal vibration, 118.
Lost work, 94.

MAGNETIC axis, 187.
—— curves, 182.
—— induction, 180.
—— meridian, 187.
—— moments, 188.
Magnetism, effect of earth, 189.
—— terrestrial, 186.
Magnetizing iron and steel, 183.
Magnets, properties of, 180.
Magnification, 146.
Manometer, 63.
Mass, measurement of, 49.
—— unit of, 11, 13, 17.
Matter, 14.
—— properties of, 15.
Maximum density of water, 166.
Measurement of angles, 33.
—— of area, 35.
—— of curved lines, 32.
—— of force, 77.
—— of length, 18, 21, 28.
—— of mass, 49.
—— of volume, 45.
—— of work, 92.
Mechanical advantage, 94, 97.
Melting-point of beeswax, 171.
—— of paraffin, 171.
Mercurial thermometer, 163.
Mirrors, 143.
—— concave, 148.
—— inclined, 143.
—— principal focus of, 148.
Moment of a force, 83.
Momentum, 79.
—— unit of, 79.
Monochord, 124.
Multiple images, 143.
Musical sounds, 125.

NEWTON'S laws, 78, 79.
Nicholson's hydrometer, 69.
Nut, 31.

OHM'S law, 215.

PARALLEL forces, 82.
Parallelogram, area of, 37.
Pendulum, 110.
—— compound, 111.
Photometer, 130.
—— Bunsen's, 133.
—— Rumford's, 132.
Pipette, 49.
Pitch of a note, 125.
—— of screw, 31.
Planimeter, 40.
Polarization, 205.
Porosity, 15.
Positive and negative electrification, 192
Potential energy, 92.
—— difference of, 199.
Power, 94.
Principal focus of lens, 145.
—— of mirror, 148.
Principle of Archimedes, 58.
—— of work, 94
Prism, volume of, 46.
Proof planes, 195.
Properties of air, 71.
—— of magnets, 180.
—— of matter, 15.
Protractor, 34.
Pulley, 98.
—— blocks, 101.
Pyramid, volume of, 48.

QUANTITY and current, 216.

RADIANT energy, 130.
Rectangle, area of, 25.
Reflection, 136.
—— total, 140.
Refraction, 137.
—— angle of, 138.
—— index of, 139.
—— through glass prism, 144.
Refractive index of glass, 141.
—— of water, 140.
Representation of a force, 77.
Resistance to flow of electricity, 214.
Resonance, 128.
Resultant force, 81.
Rumford's photometer, 132.

SAVART'S wheel, 120.
Scale, copy of, 20.
—— diagonal, 23.
Scales, 18.
Screw, 30, 103.
—— pitch of, 31.

INDEX.

Screw-gauge, 26.
— -jack, 103.
— -press, 103.
— square thread, 31.
— V-thread, 31.
Shadow photometer, 132.
Single movable pulley, 101.
Siphon, 66.
— barometer, 74.
Siren disc, 126.
Size of images, 140.
Slide-callipers, 24.
Smee's cell, 205.
Solenoid, 211.
Solids, expansion of, 155.
Sound, musical, 125.
— waves of, 117.
Space described, 80.
Specific gravity, 53.
— bottle, 56.
— by means of balance, 62.
— by means of U-tube, 64, 65.
— of liquids, 62.
— of solids, 59, 60.
Specific heat, 168.
— of glass, 170.
— of lead, 169.
Spectrum, 150.
Sphere, volume of, 45, 47.
Square, area of, 35.
Squared paper, use of, 37.
Standard yard, 153.
Steam, latent heat of, 173.
Superficial expansion, 155.

Temperature, 152.
Terrestrial magnetism, 186.
Thermometer, air, 158.
— alcohol, 162.
— differential, 160.
— fixed points of a, 164.
— graduation of, 165.
— mercury, 163.
Thermometers, 162.
Time of vibration of magnet, 189.
To copy a scale, 20.
To measure circumference of a circle, 28.
— diameter of a sphere, 29.
Total reflection, 140.
— work, 95.

Transverse vibration, 124.
Triangle, area of, 35.
— of forces, 82.

Unit of area, 11, 12.
— of force, 79.
— of length, 10, 12, 17.
— of mass, 11, 13.
— of momentum, 79.
— of volume, 10, 12, 17.
— of work, 92.
Useful constants, table of, 222.
— work, 94.
U-tube, 63, 65.

Velocity, 80.
— average, 80.
— variable, 80.
Vernier, 21.
Vibration, 110.
— longitudinal, 118.
— of thin lath, 121.
— transverse, 117, 124.
Voltaic cells, 205.
Volume, measurement of, 45.
— of cone, 47.
— of cube, 47.
— of cylinder, 46.
— of liquid, 49.
— of prism, 46.
— of pyramid, 48.
— of sphere, 45, 47.
— unit of, 10, 12, 17.

Water, latent heat of, 172.
— maximum density of, 166.
— refractive index of, 140.
Wave length, 121.
Waves, ether, 130.
— of sound, 117.
Wheatstone's bridge, 217.
Wire-gauge, 25, 27.
Work, 92.
— lost, 94.
— measurement of, 92.
— principle of, 94.
— total, 95.
— useful, 94.

Yard standard, 153.

THE END.

www.ingramcontent.com/pod-product-compliance
Lightning Source LLC
Chambersburg PA
CBHW021831230426
43669CB00008B/940